Daniel Praia Portela de Aguiar

Influence of hydro-edaphic factors on the igapó tree community

Daniel Praia Portela de Aguiar

Influence of hydro-edaphic factors on the igapó tree community

Jaú National Park, Central Amazonia, Brazil

ScienciaScripts

Imprint

Cover image: www.ingimage.com

This book is a translation from the original published under ISBN 978-613-9-65753-7.

Publisher:
Sciencia Scripts
is a trademark of
Dodo Books Indian Ocean Ltd. and OmniScriptum S.R.L publishing group

120 High Road, East Finchley, London, N2 9ED, United Kingdom
Str. Armeneasca 28/1, office 1, Chisinau MD-2012, Republic of Moldova, Europe
Printed at: see last page
ISBN: 978-620-7-79942-8

CONTENTS

1. INTRODUCTION

1.1. Amazon floodplains

Junk and Piedade (2010) define floodplain forests as all forest types subject to irregular, seasonal or long-lasting flooding. McCormick (1979) refers to these forests as lowland terrestrial ecotones, which derive their high hydrological indices and alluvial soils from the drainage and erosion of adjacent uplands, on the one hand, or from periodic flooding from aquatic ecosystems, on the other. Ilhardt *et al.* (2000), in turn, propose a functional definition, establishing that these areas are ecotones of three-dimensional interaction that include terrestrial and aquatic ecosystems, extending below the water surface, through the floodplain to the slopes adjacent to terrestrial ecosystems, and along watercourses at different amplitudes.

Riparian wetlands can also be approached on three scales: small-scale, characterizing the banks of bodies of water that become periodically flooded for a few tens of meters; medium-scale, consisting of bands of vegetation; and large-scale, forming extensive floodplains tens of kilometers long along large rivers. In the latter case, the complexity of these environments is such that scientists have given them the status of specific ecosystems (Junk, 1980; Odum, 1981; Mitsch & Gosselink, 2000).

The development of extensive floodplains along the lower and middle reaches of the Amazon River and its major tributaries was strongly influenced by a period of climate change during the Pleistocene. Severe glacial periods led to a lowering of the sea level by more than 130 m, increasing the slope of the rivers and promoting deep erosion in their valleys.

This affected the lower reaches of the tributaries of the Amazon River in the north and south up to the edges of the crystalline shields, as well as the main channel of this river up to 2500 km west inside the continent. During the interglacial periods, the sea level rose and dammed the rivers in their valleys. Over the following millennia, the rivers filled their valleys with sediment, forming extensive floodplains (Junk & Piedade, 2010).

Approximately 600,000 km^2 of the Amazon Basin are classified as wetlands influenced by seasonal flooding (Melack & Hess, 2010). Of this total, between 60.7 and 99 % is made up of tropical forests that extend along the floodplains of the lower reaches of all the major Amazonian rivers (Klinge *et al.*, 1990; Junk, 1993; Melack & Hess, 2010) and the southern rivers of the Orinoco Basin (Godoy *et al.,* 1999; Rosales *et al.,* 2001, 2002), classified, according to the system proposed by Veloso *et al.* (1991) for the vegetation of Brazil, as Alluvial Dense Ombrophilous Forests. Narrow strips of riparian forest also occur along small streams and rivers located in relatively better-drained lowlands. Their small lateral extent has led to underestimates of the total area occupied by this type of forest, and they are often included in the domain of terra firme forests (Junk & Piedade, 2010), under the generic name of *lowland forests* (Ribeiro *et al.*, 1999). In addition to these, other wetland phytophysiognomies typical of the Brazilian Amazon include large areas of insufficiently drained forests on white sand in the upper Rio Negro (Klinge & Medina, 1979; Bongers *et al.*, 1985), constituting the campinaranas (also in the dryland domain) (Veloso *et al.*, 1991); marshes dominated by swamps (Klinge & Medina, 1979; Bongers *et al.,* 1985), 1991); swamps dominated by the *Mauritiaflexuosa* palm (buritizais); wet savannah areas; mangroves (Junk & Piedade, 2010); as well as tidal flats (Prance, 1979; Junk *et al.* 2011; Junk *et al.,* 2014).

The first approach to classifying the flooded forests of the Amazon according to hydrological and hydrochemical parameters was carried out by Prance (1979), with the aim of accurately defining the terminology used for the different types of forest, since the local terms *vàrzea* and *igapó* were used indiscriminately in scientific literature (e.g.: Richards, 1952; Ducke & Black, 1953; Moreira, 1970). Sioli (1956), however, had *already* correlated the hydrochemical conditions in Amazonian rivers to the geology and geomorphology of their catchment areas.

White water rivers, such as the Amazon and the Purus, originate in the Andean or pre-Andean region, from where they transport large volumes of nutrient-rich sediments. These nutrients are deposited in large floodplains, locally called vàrzeas, where communities of terrestrial and aquatic herbaceous plants and flooded forests are established, covering an area of approximately 275,000 km^2 (Junk, 1989; Melack & Hess, 2010).

On the other hand, dark-water rivers such as the Rio Negro and the Rio Jaù drain large areas of white sand in the Paleozoic and/or Precambrian shields of the Guianas and in the interior of the basin (Junk, 1989). Their waters contain a low volume of suspended material and high concentrations of humic and fulvic acids, which give them a reddish-brown color. The floodplains of these rivers are of low fertility, locally called igapós, covered by a slow-growing forest (Furch & Junk, 1997). In this environment, aquatic macrophytes are scarce and many whitewater tree species are absent (Junk & Piedade, 1997). In fact, Wittmann *et al.* (2012) showed low floristic similarity between the vàrzea and the igapó (21 %). Due to their low fertility, igapós are consequently low productivity environments, reflecting, among other things, on tree growth (Junk *et al.*, 2015). Schongart *et al.* (2005), for example, comparing growth patterns between igapó and floodplain populations of the species *Macrolobium acaciifolium,* recorded that, under the same flooding regime, the species showed a significantly lower average radial increment in the igapó (1.52 ± 0.38 mm) than in the floodplain (2.66 ± 0.67 mm). The igapó trees, however, had a higher average age (268 ± 118 years) than the vàrzea (135 ± 27 years), indicating that in the igapó the trees take longer to reach the same diameter as the vàrzea trees.

The clear water rivers, such as the Tapajós and the Tocantins, have their catchment areas in the Brazilian Central Plateau. Their waters are transparent and greenish, with a low volume of sediment and dissolved solids, and the floodplains have intermediate fertility and are also classified as igapós. The igapós, in general (dark and light waters), cover an area of approximately 180,000 km^2 in the Amazon Basin (Junk, 1989; Melack & Hess, 2010).

A third type of Amazonian floodplain forest, first described by Irion (1976), is the paleo-wetland forest. This type of forest is found on ancient alluvial deposits of Andean origin abandoned by white-water rivers and currently flooded by dark-water rivers (e.g. Jutai, Tefé, Coari rivers), which carry intermediate levels of long-deposited Andean sediments (Irion *et al.*, 2010; Junk *et al.*, 2011). These sediments form substrates of lower fertility than those of the vàrzea, but higher than those of the igapó (Irion, 1978; Schubart, 1983). Paleo-variezas cover an area of 125,000 km^2 in the Amazon Basin (Junk *et al.*, 2011; Assis *et al.,* 2015).

In a study by Costanza *et al.* (1997), which estimated the monetary value of the services offered by different ecosystems on a global scale, floodplains obtained the second highest value per hectare per year, behind estuaries. The value obtained by De Groot *et al.* (2012), on the other hand, was the fourth highest, behind the values obtained for coral reefs, coastal wetlands and coastal systems (in the latter case, considering estuaries, continental shelves and algae beds together). Globally, without taking into account area units, floodplains obtained the sixth highest value, behind the open sea, continental shelves, estuaries, tropical forests and algae banks (Costanza, *et al.* 1997). The floodplain forests of the Amazon, within the context of the continental wetlands, are of great economic and ecological importance in the region, especially their use as a source of timber and non-timber products; their role as a natural barrier to the flow of rivers, protecting the floodplain from erosion; its carbon sequestration activity and microclimate regulation; its use as a habitat and food source for many animals, including economically important fish species; its use as a grazing area for cattle; its genetic stock of species adapted to flooding; among others (Junk & Piedade, 2010; Junk *et al.*, 2014).

1.2. Diversity and floristics of Amazonian floodplain forests

The richness of tree species in the floodplains of the Amazon is lower than that of the adjacent terra firme (Prance, 1979; Balslev *et al.*, 1987). Recent studies, however, have shown that, among floodplain forests, floodplain forests are the richest in species worldwide (Wittmann *et al.*, 2006; Wittmann et al., 2012; Assis *et al.,* 2015), which can be explained by the high diversity of habitats, whose establishment depends mainly on geo-hydrological factors; and by the relatively stable geomorphological conditions in the Amazon Basin over millions of years.

In contrast to the terra firme forests of the Amazon, for which databases published up to 2006 report more than 250,000 individuals in an inventoried area of more than 700 ha (Ter Steege *et al.*, 2006), the number of trees inventoried in the swamp forests is small. In floodplain forests, floristic inventories carried out up to the same year had covered an area of less than 70 ha and approximately 40,000 individuals (Wittmann *et al.*, 2006), 2006), while the databases for igapó contained inventories of less than 20 ha and approximately 11,000 individuals (Rodrigues, 1961; Keel & Prance, 1979; Revilla, 1981; Piedade, 1985; Campbell *et al.*, 1986; Ferreira, 1991; Ferreira & Prance, 1998; Haugaasen & Peres, 2006; Inuma, 2006).

Kubitzki (1989), based on Prance's (1979) classification, exposed the floristic differences between floodplain and igapó, taking into account differences in nutrient levels. Kubitzki (1989) also pointed to strong floristic relationships between floodplain forests and adjacent terra firme forests. The flora of the igapós is also closely related to that of the campinaranas, where, in certain circumstances, the periodically high water table or even surface flooding require the trees to adapt accordingly. Many of these species, such as *Ruizterania retusa*, *Panopsis rubescens* and *Humiria balsamifera,* are also found in the higher parts of the igapós. Other genera, however, such as *Leopoldinia* and *Lophantera*, are restricted to periodically flooded areas. Targhetta (2012), on the other hand, comparing the floristics of the campinarana to that of the igapó on the Uatuma River, found low similarity between these two environments, with only four species common to both. The author attributes this opposite result to potential differences in floristic composition between igapós in different locations, under the influence of different environmental factors. On the other hand, the comparison between the two studies also shows that the greatest similarity between the igapó and the campinarana is at the family and genus level, while this similarity decreases at the specific level,

Assis *et al.* (2015), in a study on the diversity and floristic composition of the paleo-varzea, showed that it contains elements of both the vàrzea and the igapó, but is floristically more similar to the vàrzea. However, it points to the tendency that, as the sedimentation process slows down, the palaeovarzeas will become floristically more similar to igapós, with a gradual replacement by species more common to the latter.

1.3. The role of hydro-edaphic factors in floristic diversity and composition

The dynamics of the Amazon's forests and other floodplain ecosystems are strongly influenced by periodic flooding, the *Flood* Pulse *Concept* (FPC) of which was developed by Junk *et al.* (1989) to describe the processes that occur on the floodplains of large Amazonian rivers. This concept predicts that the flood pulse is the force that directs floodplain systems, both conditioning the occurrence and distribution of plants and animals and determining the characteristics of their life histories, affecting primary and secondary productivity and influencing decomposition and nutrient cycles in water and soils.

In the Amazon, the flood pulse stems mainly from the seasonality of rainfall in the region where the rivers originate and from the Andean thaw. It is therefore classified as monomodal and predictable, with the ebb, dry, flood and flood phases occurring only once in each cycle, at specific times throughout the year. This predictability makes it easier for organisms to adapt to the changes between the aquatic and terrestrial phases, increasing their ability to make efficient use of the periodically available resources (Junk *et al.*, 1989).

The trees that establish themselves on these floodplains make use of different adaptations to cope with the periodic oxygen deficit and the associated shortening of the growing season. The high sedimentation rate near the whitewater river channel, the scarcity of nutrients in the dark and clear water floodplains and the anoxic conditions of the rhizosphere in depressions with stagnant water accentuate the complex phytoecological interactions to which the trees are subject (Wittmann *et al.*, 2010a). Among the main adaptations are the synchronization of wood growth during the terrestrial phase (Worbes, 1989; Schongart *et al.*, 2005), synchronization of reproductive phenology with the aquatic phase and fruits adapted to ichthyochory (Pires & Prance, 1985; Schongart *et al.*, 2002; Oliveira Wittmann et *al.*, 2007), the formation of adventitious roots and the increase in root biomass during the period of high water (Worbes, 1989; Wittmann & Parolin, 2005), leaf loss and/or reduced photosynthesis during flooding (Parolin *et al.* 2004a), anaerobic metabolism (Schlüter *et al.*, 1993), the formation of layers

oxygenated around submerged roots and the incorporation of suberin into the rhizodermis (De Simone *et al.*, 2002), high sugar reserves in the roots (Ferreita *et* al., 2009) and the persistence of fully functional leaves in evergreen species (Waldhoff *et al.*, 2002; Parolin, 2009).

In their natural environment, trees are distributed along the flood gradient in a characteristic pattern that is determined by abiotic and biotic factors. The main abiotic factors are the duration, depth, frequency, synchronization and shape of the flood pulse, current velocity, sedimentation rate, sediment quality, erosion and water stress during low water. The main biotic factors are competition, organic matter content and oxygen concentrations in the soil, seed dispersal and seedling establishment. Large trees survive flooding better than small trees, as the duration of flooding is generally correlated with the depth of flooding, and many species cannot withstand total submergence during the growing season (Oliveira Wittmann *et al.*, 2010).

Some studies have reported lower species richness and differential distribution of species along the flood gradient from intensely flooded areas to less flooded areas, attributing this to different levels of tolerance of species to flooding and to the characteristics of the topography and soil (e.g.: Takeuchi, 1962; Keel & Prance, 1979; Adis, 1984; Campbell *et al.,* 1992; Ayres, 1993). In studies carried out near the cities of Tefé and Manaus (Wittmann, 2001; Wittmann *et al.*, 2002), the varzea forests were differentiated, according to the flooding gradient, into two basic types: low varzea forests, whose average flooding reaches heights of between 3.0-7.5 m, which corresponds to a flooding period of between 50 and 230 days per year; and high varzea forests, whose average flooding reaches heights of less than 3.0 m, with an average flooding period of less than 50 days per year. The results obtained make it clear that there is a zonation along the flood gradient, reflected in characteristic forest typologies, with only a few species (17.5%) occurring along the entire gradient and many species restricted to small topographic ranges (Parolin *et al.*, 2002).

Due to the intense process of erosion and sedimentation that promotes the continuous development and disappearance of sites for the establishment of vegetation over time, varzea forests can be classified in more detail according to a successional sequence (*sensu* Ricklefs, 1990), where there is a continuous increase in species diversity and a decrease in the number of individuals per species. Taking into account floristic surveys and the nomenclature adopted in various works (e.g.: Worbes *et al.*, 1992; Ayres, 1993; Schongart, 2003; Wittmann *et al.,* 2004), Wittmann *et al.* (2010a) presented the delineation of the following typologies for lowland rainforest: primary succession along rivers and river channels, primary succession in lakes, early secondary succession along rivers, lowland late secondary stage, lowland late succession, highland rainforest (climax stage) and chavascal.

The vast majority of studies on Amazonian floodplain forests have focused on floodplain forests. Consequently, much more knowledge about their botany, taxonomy, phytogeography, ecology and physiology is available than for igapó forests (Wittmann *et al.*, 2010a). Although, in terms of hydrogeomorphological dynamics, igapós are relatively stable and low-productivity habitats compared to floodplains (Sioli, 1975; Junk, 1989), some studies have already shown that the variation in the flood gradient exerts a certain influence on floristic diversity and composition in these environments. These include the work by Ferreira (2000), conducted in parallel on the Jaù and Tarumâ-Mirim rivers (AM), and the study by Ferreira & Parolin (2011) on the Tarumâ-Mirim river. Montero *et al.* (2014), on the other hand, conducted a study to describe the floristic variation of the igapó along the course of the Rio Negro, without taking into account the flood gradient. The author observed abrupt variations in species composition between different sections of the river and a higher diversity in the sections located in the lower course, which are based on more recent geological formations and are therefore richer in sediments and nutrients, due to the influence of the Branco River.

Another important abiotic factor that influences the distribution of species between different environments is the soil. Among the various soil properties, texture (the proportion between the sand, silt and clay fractions) intensely interferes with water and nutrient retention, limiting the establishment of certain plant species depending on the textural proportion (Vieira & Santos, 1987; Resende *et al.,* 2007). Several authors have suggested or taken into account the role of edaphic factors in structuring the igapó tree community (e.g. Ferreira, 2000; Hamaguchi, 2009; Targhetta, 2012). Worbes (1986, 1997) observed that igapó generally has areas on sandy soil and others on clay soil, reflecting marked differences in species composition between sites with different soil textures (Junk *et al.,* 2015).

No study to date, however, has found zonation patterns along the hydro-edaphic gradient of the igapó that would allow forest typologies to be delimited, and there is a need for studies using more systematic methods in this regard.

2. OBJECTIVES

2.1. General objective

To analyze the influence of the flooding gradient and the edaphic gradient on the composition, richness, diversity and structure of the igapó tree component.

2.2. Specific objectives

- To analyze the influence of flood duration and edaphic variables on the composition, richness, diversity and structure of the igapó tree component;
- Describe the zonation of species along environmental gradients.

3. MATERIAL AND METHODS

3.1. Description of the study area

This work was carried out in the Jaù National Park (PARNA Jaù), located approximately 200 km northwest of Manaus, in the state of Amazonas, covering the watersheds of the Jaù and Unini rivers, tributaries of the right bank of the Negro river. The park's boundaries were drawn along the banks of the main rivers, with the northern boundary determined by the Unini River and the southern boundary by the Carabinani River (Fundaçao Vitória Amazônica, 1998).

The rivers that delimit the park have dark waters and originate in low-lying terrain. Rainfall is high from January to May, with the highest river levels in June and July. The period of lowest rainfall is between July and September, and the lowest river levels are recorded in October and November (Fundaçao Vitória Amazônica, 2008).

PARNA Jaù is located on a mosaic of geological formations of different ages: Prosperança (Cambrian-Ordovician), occupying 17% of the area; Trombetas (Silurian), 8%; Barreiras/Alter do Chao (Cenozoic), 3%; Solimoes (Tertiary and Pleistocene), 65%; as well as Holocene alluvial deposits (Quaternary), corresponding to 7% of the area (Radambrasil, 1978; Schubart, 1977).

The park's vegetation is made up of 75.3% terra firme forests (Lowland Dense Ombrophilous Forest), 1.93% submontane forests (Submontane Dense Ombrophilous Forest), 13.11% igapó forests (Alluvial Dense Ombrophilous Forest), 0.9% grasslands and 8.76% campinaranas (Radambrasil, 1978; Veloso *et al.*, 1991; Fundaçao Vitória Amazônica, 2008).

3.2. Data collection

Ten plots (P1 to P10) measuring 50 x 200 m (1 ha) were marked out in the igapó forest of PARNA Jaù, approximately parallel to the river, belonging to the Long-term Ecological Project (PELD) of the Research Group on Ecology, Monitoring and Sustainable Use of Wetlands (MAUA) (Appendix A). The plots were established in places subject to different average flood heights, identified by observing the obvious marks on the tree trunks before the surveys began. The aim was to cover the widest possible range of the flood gradient, without, however, typifying the plots on the basis of topographical or physiognomic classifications.

Each plot (Figure 1) was divided into 16 subplots measuring 25 x 25 m, named after a letter in the sequence A to P. Each set of four sub-plots was called a quadrant, and they were named according to the circumscribed sub-plots (e.g. quadrant ABCD). All the vertices of the plots were geo-referenced in order to record the exact location of the plots and facilitate later work.

Data was collected at the end of October 2013 and in March 2014, when the Jaù River and its tributaries were at low levels and the igapó forest was not flooded. Based on the maximum level reached by the Jaù River in 2013, the mark of which could be seen on the trunks of the igapó trees, the average annual flooding period for each individual was calculated (in days per year^{-1}), as well as the average height of flooding (in meters) to which each individual was subjected. The aim was to obtain average values for the duration and height of flooding for each species. The hydrological data was obtained from the website of the National Water Agency through the Hidroweb system (Hidroweb, 2015), using as a reference the daily levels of the Jaù River obtained at the Seringalzinho Station (code 14876000), for the period from 01/10/2005 to 01/10/2014, corresponding to 10 years of records.

All arborescent trees and palms with a DBH $\geq$ 10 cm present in each plot were taken into account. Identifications were previously made in the field with the help of a parataxonomist, but botanical material was collected from some of the individuals in order to help specialists determine doubtful identities and/or by comparing them with exsiccates held in the herbarium of the National Institute for Amazonian Research (INPA). To this end, the samples were obtained with pruning shears, pressed and then herborized. The phylogenetic system adopted was APG III (APG III, 2009).

All the individuals were marked with numbered aluminum plates and the DBH was measured using a diametric tape in centimeters. Individuals with sapopemas had their DBH measured above the sapopemas. Individuals with fluted trunks had their DBH taken without taking into account the indentations in the trunk.

For the edaphic analyses, samples were taken approximately 20 cm deep from the central point of each quadrant (Figure 1), totaling four samples per plot. The material was sent to the EMBRAPA (Empresa Brasileira de Pesquisa Agropecuaria) - Amazônia Ocidental laboratory and analyzed according to the specific methodology. Sixteen variables were taken into account, the physical variables being sand, silt and clay and the chemical variables being C, N, P, K^+ , Ca^2 +, Mg^2 +, Al^{3+} , Fe, Zn^+ , Mn^{2+} , Cu, organic matter and pH.

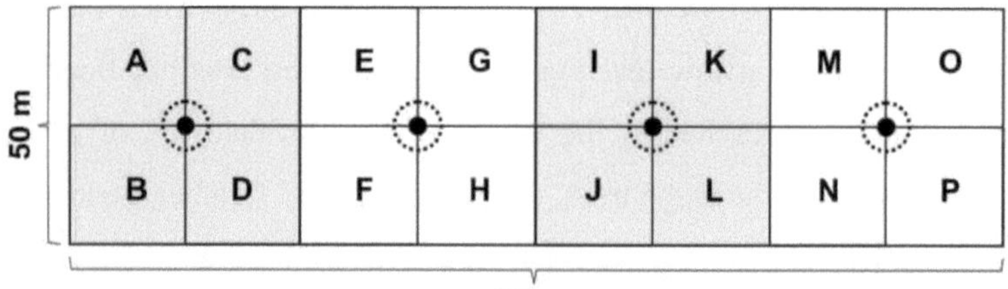

Figure 1 - Plot layout used. The quadrants are made up of groups of four sub-plots. The circles indicate the soil sampling points within the quadrant.

3.3. Data analysis

After determining the identity of the species, the richness (*S) of* the set of plots was calculated, as well as the richness per plot and per quadrant. In order to demonstrate sampling sufficiency, a species accumulation curve was drawn up based on the inclusion of new individuals throughout the inventory (Colwell & Coddington, 1994).

The diversity of the set of plots, the diversity per plot and per quadrant were obtained using Fisher's coefficient (*Fisher*'s *α*) (Fisher *et al.*, 1943), as it is a coefficient relatively unaffected by variations in sample size and completely independent of it if N>1000 (Taylor, 1978). In addition to this index, the Shannon diversity index (H') (Krebs, 1999) was calculated for the set of plots and per plot, in order to allow comparisons with other floristic surveys. Both coefficients were calculated using the Past 2.16 software (Hammer *et al.*, 2001).

Based on the DBH data and the species count and identification, the following phytosociological parameters (Curtis & McIntosh, 1951) were calculated at species and family level using the Fitopac 2.1.2 software (Shepperd, 2010):

- *Relative Density* (RD_i) = [number of individuals of species/family *i* sampled in the plot (*Ni*) / total number of individuals sampled in the plot ($\sum Ni$)] x 100
- *Relative Dominance* (DoR_i) = [basal area of all individuals of species/family *i* sampled in the plot (Ab_i) / basal area of all individuals sampled in the plot ($\sum Abi$)] x 100;
- *Relative Frequency* (RF_i) = [Absolute frequency of species/family *i* (AF_i) / sum of the absolute frequencies of all species/families sampled in the plot ($\sum FAi$)] x 100, where: [Absolute frequency of species *i* (AF_i) = number of plots in which species/family *i* occurred (N_{Pi}) / total number of plots ($\sum N_{Pi}$) x 100];
- *Importance Value Index* (IVI) = $DR_i + DoR_i + FR_i$.

For the statistical analysis of hydro-edaphic factors in floristics and structure, the values of both the predictor variables (duration of flooding and edaphic variables) and the response variables (richness, diversity, basal area (cm^2) and density (individuals ha^{-1})) per quadrant were considered. This scale of approach was chosen because the greatest detail obtained when collecting edaphic data was at quadrant level. The values corresponding to the richness, diversity, composition, basal area and density of the set of plots and of each plot were presented in this study in order to describe general aspects of the area in question and to allow comparison with other studies, and were not analyzed statistically.

To analyze the influence of the flooding gradient on richness, diversity, basal area and density, Simple Linear Regressions were performed between the values of these variables and the value of the average duration of flooding per quadrant. To check the effect of the flooding gradient on species composition, a Non-Metric Multidimensional Scaling (NMDS) was carried out with the abundance values of individuals per species within each quadrant. A Simple Linear Regression was then carried out between the first axis generated by the NMDS and the value of the average duration of flooding per quadrant, in order to verify the influence of flooding on floristic composition.

In order to demonstrate the influence of the flooding gradient on the variation in floristic composition, a table was drawn up in which the species, excluding those represented by *singletons* (Novotny & Basset, 2000), were organized and classified according to their level of tolerance to flooding. A classification of flooding was proposed, according to the intervals considered, as *low* (0 to 3 m), *intermediate* (3 to 6 m), *high* (6 to 9) and *very high* (10 to greater than 11). Based on the greater or lesser concentration of individuals per species within each flood class, the species were classified as being of *low*, *intermediate*, *high*, *very high* and *irregular* tolerance.

In order to provide a better visualization of the distribution along the gradient, a graph was drawn up showing the relative abundance of the 60 most abundant species (minimum *number* = 22) along the gradient. For both the table and the graph, the flood values per individual were taken into account.

To analyze the influence of the edaphic factors, a PCA (Principal Component Analysis) was carried out with the values of the physico-chemical variables obtained for each quadrant. The data was standardized by dividing each value per quadrant by the maximum value of each variable. Simple linear regressions were then performed between the values for richness, diversity, basal area and density per quadrant and the first axis obtained from the PCA. To check the effect of the edaphic gradient on species composition, a Mantel test was carried out between the matrices generated by NMDS and PCA.

Plot 8 (P8) was excluded from the statistical analysis, as it was not possible to collect soil samples due to its surface being submerged at the time of the inventory.

All statistical analyses were carried out in the R software (R Development Core Team, 2011), using the Vegan package (Oksanen *et al.*, 2009).

4. RESULTS

4.1. General vegetation structure

The histograms relating the relative density of individuals according to diameter class for each plot (Appendix A) generally showed an inverted J shape, with most of the individuals inventoried concentrated in the smaller diameter classes. Plot 8 (P8), however, did not follow this pattern, with most of the individuals concentrated in the 30 - 40 cm (23.4%) and 40 - 50 cm (19.6%) classes.

Considering the set of plots, the histogram showed an inverted J shape (Figure 2), with 58.7% of the individuals concentrated in the 10 - 20 cm class.

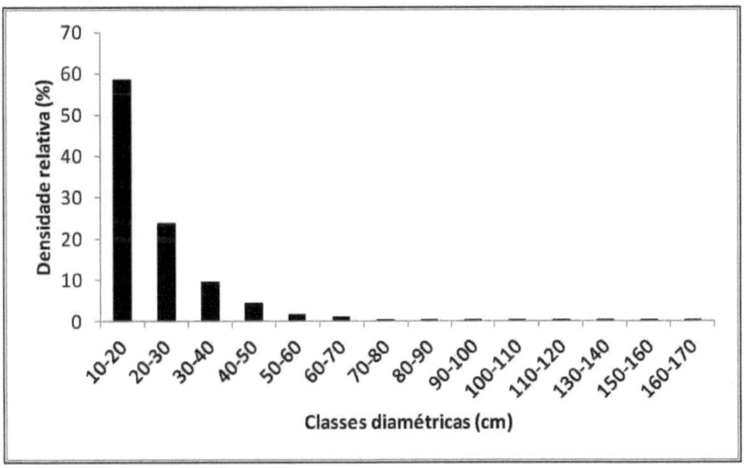

Figure 2 - Relative density of individuals according to diametric classes considering the set of 1 ha plots.

4.2. General aspects of richness, diversity and floristic composition

A total of 6992 individuals were recorded. The richness (S) obtained after the inventory of the 10 plots, which totaled 10 ha, was 193 species, divided between 115 genera and 41 families (Appendix B).

The species accumulation curve (Figure 3) showed no tendency to form an asymptote.

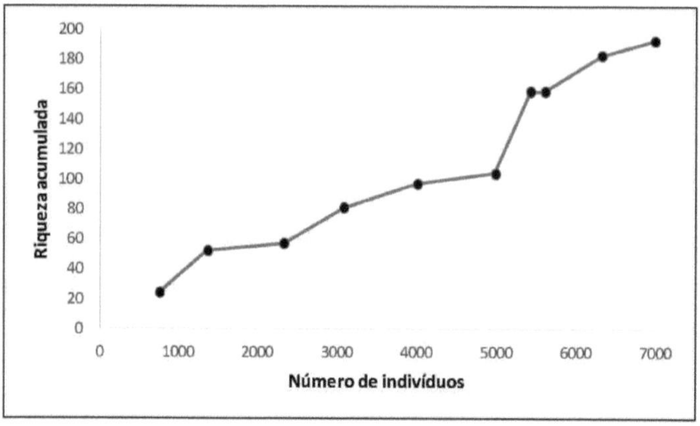

Figure 3 - Species accumulation curve based on the inclusion of new individuals.

The plot with the highest abundance was P6 (977 individuals), the plot with the highest number of families was P9 (29 families) and the richest in species was P7 (74 species). Plot 8 (P8) had the lowest abundance and was the least rich in families and species (Table 1). Species richness per hectare ranged from 3 species in P8 to 74 species ha^{-1} in P7. A total of 31 *singletons* were recorded, with P7 having the highest abundance (19).

Fisher's α between plots ranged from 0.51 (P8) to 25.58 (P7). The same plots had the lowest and highest values for the Shannon index respectively (0.12 and 3.69 respectively) (Table 1). Considering the set of plots, the values for Fisher's α and the Shannon index obtained were 39.34 and 4.19, respectively.

Table 1 - Abundance, richness and diversity values per plot and overall.

Portion	Individuals	Families	Species	Fisher's α	Shannon (H')
1	758	14	24	4,72	2,27
2	607	22	44	10,9	2,83
3	962	14	26	4,92	2,33
4	755	26	54	13,32	3,34
5	928	27	50	11,31	3,02
6	977	24	52	11,73	2,93
7	436	26	74	25,58	3,69
8	184	3	3	0,51	0,12
9	718	29	61	15,92	2,99

10	667	28	67	18,56	3,68
Set of plots	6992	41	193	39,34	4,19

4.3. Phytosociological characterization

The ten most abundant species in the plots were, in descending order: *Pouteria elegans* (499), *Amanoa oblongifolia* (497), *Macrolobium acaciifolium (*306), *Elvasia quinqueloba* (304), *Burdachia* sp. (269), *Eschweilera tenuifolia* (257), *Duroia* cf. *velutina (*254), *Swartzia polyphylla (*239), *Hevea spruceana* (179) *and Erythroxylum spruceanum (*166). Together, these species accounted for 42.5% of the total abundance. Two species of the genus *Tachigali* (*T. hypoleuca* and *T. rigida*) were also among the most abundant, but the number of individuals of each species could not be obtained due to the difficulty in differentiating the two species in the field most of the time, mainly because it was impossible to take samples of all the individuals. It is estimated that *T. hypoleuca*, however, is more abundant and more frequent than *T. rigida*.

The species with the ten highest values for Relative Density, Relative Frequency, Relative Dominance and Importance Value are summarized below (Table 2). The species with the highest Relative Density was *Pouteria elegans* (7.14%), *Ternstroemia candolleana* was the most frequent (1.98%), *Eschweilera tenuifolia* the most dominant (9.09%) and *Macrolobium acaciifolium* the most important (14.81).

Table 2 - Species with the ten highest values for the phytosociological parameters considered.

Relative density		**Relative frequency**	
Species	**Value (%)**	**Species**	**V value (%)**
Pouteria elegans	7,14	*Ternstroemia candolleana*	1,98
Amanoa oblongifolia	7,11	*Pouteria elegans*	1,76
Macrolobium acaciifolium	4,38	*Macrolobium acaciifolium*	1,76
Elvasia quinqueloba	4,35	*Maprounea amazonica*	1,76
Burdachia sp.	3,85	*Burdachia* sp.	1,54
Eschweilera tenuifolia	3,68	*Swartzia polyphylla*	1,54
Duroia cf.*velutina*	3,63	*Aniba affinis*	1,54
Swartzia polyphylla	3,42	*Amanoa oblongifolia*	1,32
Hevea spruceana	2,56	*Eschweilera tenuifolia*	1,32
Erythroxylum spruceanum	2,37	*Duroia* cf.*velutina*	1,32

Relative Dominance		Importance (IVI)	
Species	Value (%)	Species	Value
Eschweilera tenuifolia	9,09	*Macrolobium acaciifolium*	14,81
Macrolobium acaciifolium	8,68	*Eschweilera tenuifolia*	14,09
Swartzia polyphylla	7,93	*Amanoa oblongifolia*	13,46
Aldina latifolia	6,14	*Swartzia polyphylla*	12,89
Amanoa oblongifolia	5,03	*Pouteria elegans*	12,49
Ormosia excelsa	3,71	*Aldina latifolia*	9,52
Pouteria elegans	3,59	*Burdachia* sp.	7,14
Hevea spruceana	3,28	*Elvasia quinqueloba*	7,12
Hydrochorea marginata	2,58	*Ormosia excelsa*	6,86
Micropholis humboldtiana	2,13	*Duroia* cf.*velutina*	6,85

The ten most abundant families were, in descending order: Fabaceae (1956), Sapotaceae (821), Phyllanthaceae (518), Lecythidaceae (399), Euphorbiaceae (386), Malpighiaceae (354), Ochnaceae (320), Rubiaceae (264), Chrysobalanaceae and Myrtaceae (194). Together, these families accounted for 78% of the total abundance. In terms of richness, the families with the ten highest values were: Fabaceae (39), Sapotaceae (15), Lecythidaceae (11), Chrysobalanaceae (11), Myrtaceae (11), Annonaceae (11), Euphorbiaceae (8), Malpighiaceae (7), Lauraceae (7) and Melastomataceae (7). Together, these families accounted for approximately 66% of the total richness.

The families with the ten highest values for Relative Density, Relative Frequency, Relative Dominance and Importance Value are summarized below (Table 3). Fabaceae had the highest Relative Density (27.97%), was also the most dominant family (47.24%), the most important (79.91) and was among the three most frequent, along with Lecythidaceae and Malpighiaceae (all with a Relative Frequency of 4.69%).

Table 3 - Families with the ten highest values for the phytosociological parameters considered.

Relative Density		Relative Frequency	
Family	Value (%)	Family	Value (%)
Fabaceae	27,97	Fabaceae	4,69
Sapotaceae	11,74	Lecythidaceae	4,69
Phyllanthaceae	7,41	Malpighiaceae	4,69

Lecythidaceae	5,71	Sapotaceae	4,23
Euphorbiaceae	5,52	Euphorbiaceae	4,23
Malpighiaceae	5,06	Chrysobalanaceae	4,23
Ochnaceae	4,58	Pentaphylacaceae	4,23
Rubiaceae	3,78	Myrtaceae	3,76
Chrysobalanaceae	3,49	Lauraceae	3,76
Myrtaceae	2,77	Combretaceae	3,76
Relative Dominance		**Importance (IVI)**	
Family	**Value (%)**	**Family**	**Value**
Fabaceae	47,24	Fabaceae	79,91
Lecythidaceae	10,60	Sapotaceae	23,37
Sapotaceae	7,40	Lecythidaceae	21,00
Phyllanthaceae	5,15	Phyllanthaceae	15,85
Euphorbiaceae	4,98	Euphorbiaceae	14,73
Malpighiaceae	2,33	Malpighiaceae	12,09
Combretaceae	2,21	Chrysobalanaceae	9,66
Chrysobalanaceae	1,95	Rubiaceae	9,00
Rubiaceae	1,94	Ochnaceae	8,65
Lauraceae	1,77	Lauraceae	7,75

4.4. Influence of hydro-edaphic factors

The average duration of flooding to which the trees were subjected in the study *area* ranged from 6 to 288 days per year^{-1} , corresponding to a flood height of between 0.1 and 11.1 meters. Among the study plots, the one with the highest average flooding values was P8, while P10 had the lowest values (Table 4).

Table 4 - Average flooding values per plot.

Tranche	Average flood height (m)	Average duration of flooding (days per year)$^{-1}$
1	7,72 ± 0,41	219 ± 11
2	6,31 ± 0,66	184 ± 14
3	5,32 ± 1,15	162 ± 27
4	5,55 ± 0,16	169 ± 3
5	4,94 ± 0,36	156 ± 8
6	3,28 ± 0,4	108 ± 11
7	2,57 ± 1,29	81 ± 44
8	10,82 ± 0,11	284 ± 2
9	1,55 ± 0,74	45 ± 28

10	1,24 ± 0,44	33 ± 17

The first axis of the NMDS was correlated with the duration of the flood (*$p < 0.01$; $R^2 = 0.52$*) (Figure 4A), demonstrating that the variation in composition is correlated with the flood. Flood duration was negatively correlated with richness (p < *0.01; $R^2 = 0.51$*) (Figure 4B) and diversity (*$p < 0.01$; $R^2 = 0.46$*) (Figure 4C) and positively correlated with basal area (*$p < 0.01$; $R^2 = 0.33$*) (Figure 4D). In relation to density, the correlation was positive but not significant (*$p = 0.13$; $R^2 = 0.04$*) (Figure 4E).

Around 36% of the specimens (n = 70) were classified as having predominantly low tolerance to flooding (Appendix C), 56 of which were restricted to areas under low flooding; 40% (n = 78) were classified as having intermediate tolerance; 5% (n = 10) were classified as having predominantly high tolerance, with one specimen classified as having extreme tolerance; and 2% (n = 4) as having irregular tolerance.

The graph of the relative abundance of individuals per species along the flood gradient (Appendix D) also helped to visualize the greater or lesser concentration of individuals per species along the gradient.

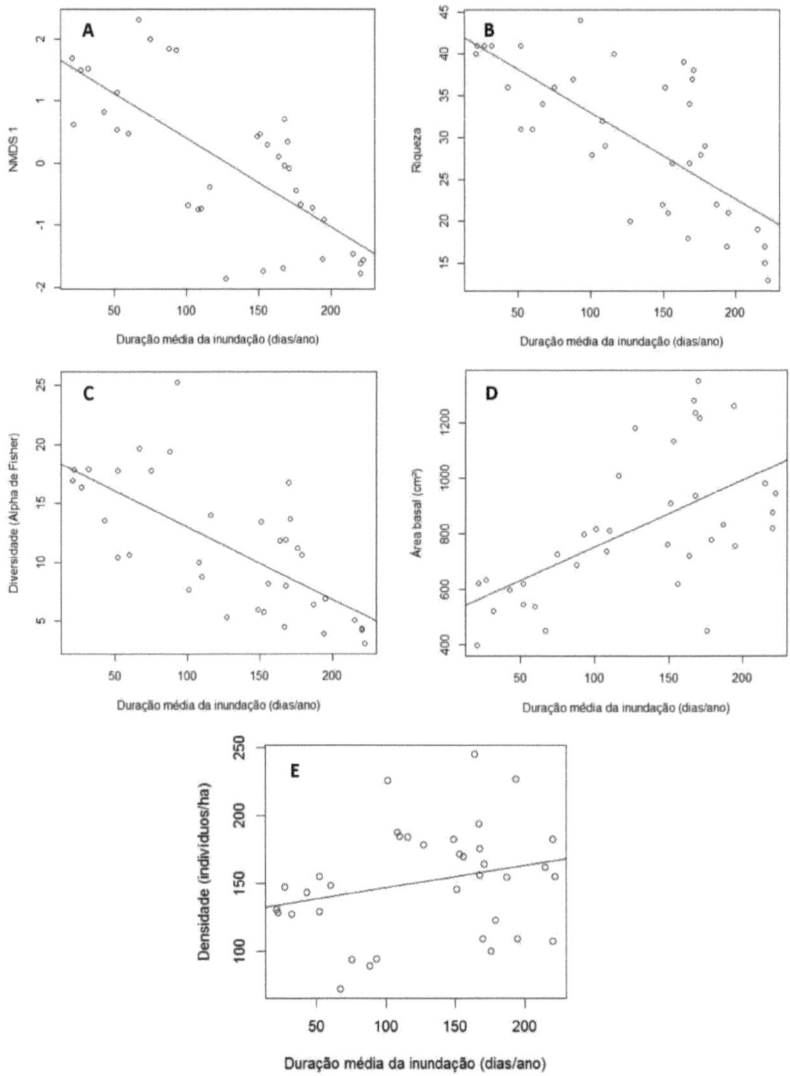

Figure 4 - Simple linear regression graphs between flood height and composition (A), richness (B), diversity (C), basal area (D) and density (E).

The first two axes of the PCA explained 42% (standard deviation: 0.63) and 26% (standard deviation: 0.49) of the edaphic variation, respectively. Considering only the first axis (Table 5), the three most positively correlated variables were carbon, organic matter and nitrogen.

The Mantel test revealed a significant correlation between species composition (NMDS) and edaphic gradient (PCA) *($p < 0.01$; $r = 0.36$)*.

The Simple Linear Regression showed an insignificant and practically non-existent correlation between the first axis of the PCA and the variables richness *($p = 0.96$; $R^2 = 0.02$)*, diversity *($p = 0.20$; $R^2 = 0.02$)* and basal area *($p = 0.92$; $R^2 = 0.03$)* and a significant but weak correlation with density *($p = 0.03$; $R^2 = 0.11$)*.

Table 5 - Coefficients (*loadings*) obtained for each of the graphical variables in the first two axes of the PCA.

Variable/Axis	PCI	PC2
pH	0,04	0,00
C	0,42	-0,04
M.O.	0,42	-0,04
N	0,37	-0,07
P	0,16	0,18
K^+	0,36	0,17
Ca^{2+}	0,17	0,18
Mg^{2+}	0,26	0,18
$Al+^3$	0,17	-0,26
Fe	-0,06	0,45
Zn^+	0,24	0,10
Mn^{2+}	0,29	0,19
Cu	0,13	0,13
sand	-0,11	0,66
silt	0,13	0,01
clay	0,19	-0,30

*O.M.: Organic matter

5. Discussion

5.1. General vegetation structure

The inverted J shape of the histogram of the distribution of relative density of individuals according to diametric classes obtained for most of the plots and for the set of plots was also observed in other studies carried out in igapó (Ferreira, 1997; Hamaguchi, 2009; Targhetta, 2012), vàrzea (Wittmann & Junk, 2003), campinarana (Stropp *et al.*, 2011; Targhetta, 2012), and on terra firme (Rankin-de-Mérona *et al.*, 1992; Oliveira & Mori, 1999; Oliveira & Amaral, 2004). This distribution generally characterizes a stock community, which is the pattern in stable tropical forests of varying age and species composition (Scolforo *et al.*, 1998), where each diametric class represents a stage in the regeneration of the fraction of the stand of the same species or of a community with a diameter greater than this class (Rollet, 1978).

P8 was the only one of the ten study plots in which the distribution according to diametric classes did not show an inverted J shape, with the majority of individuals concentrated in intermediate classes. This can be attributed to the fact that the plot in question was the one with the highest average height and duration of flooding (Table 4), which makes it difficult to establish seedlings. According to Schongart *et al.* (2005), in highly flooded areas, seedling establishment is only possible in extreme hydroclimatic events.

5.2. General aspects of richness, diversity and floristic composition

The species richness obtained in this study (193 in 10 ha) was higher than that found in other surveys carried out in igapó forests, considering DBH ≥ 10 cm as the inclusion criterion, such as those by Hamaguchi (2009), carried out in the Tupé RDS (53 species in 1.01 ha); Ferreira & Parolin (2011), on the Taruma-Mirim river (102 species in 1.44 ha); Targhetta (2012), on the Uatuma river (69 species in 3 ha). However, it was lower than the richness found by Montero *et al.* (2014) along the Negro River (243 species in 10 ha). In the latter case, where the total area inventoried was the same as in the present study, the higher richness found is attributed to the fact that the surveys carried out by that author took place in different sections along the Negro River, over a longer stretch, characterized by different flooding regimes, alluvial soil quality and fluvial geomorphology. This variation along the course of the river, according to the author, influences changes in floristic composition between different sections, contributing to an increase in richness.

The non-formation of the asymptote of the species accumulation curve in the present study suggests that the 10 plots used were not sufficient for a representative sample of the species richness of the igapó forests of PARNA Jaù, since the addition of more individuals signals a potentially significant increase in richness.

In general, the present study and others carried out in igapó forests always show lower richness than in veraison forests (Worbes, 1997; Parolin, 2000), considering DBH $\geq$ 10 cm as the inclusion criterion.

The highest species richness per hectare found in this study (74 species ha^{-1} , P7) was higher than that found in the study by Targhetta (2012), but was lower than that found by Ferreira (1997), also in the Jaù River (137 species ha^{-1}); and by Montero *et al. (*2014), in Anavilhanas (79 species ha^{-1}) (For a comparison of richness and diversity between different sites and ecosystems, see Table 6).

Although the influence of the duration of flooding was not statistically analyzed at plot level, the fact that P7 showed the greatest richness and diversity is certainly related to its greater proximity to terra firme in relation to the other plots, which means that this plot has the greatest amplitude of variation in the flood gradient among the plots (5.4 m). There is thus the possibility of establishing a greater number of species that are tolerant of different levels of flooding (Keel & Prance, 1979; Adis, 1984; Campbell *et al.*, 1992; Ayres, 1993), ranging from the elevations closest to the river to the edge of the terra firme. The proximity to terra firme also contributed to plot 7 having the highest number of *singletons*, most of which are typical of terra firme forests and rarely mentioned in other studies carried out in igapó forests, such as *Bertholletia excelsa*, *Parkia pendula* and *Simaba cedron*. Ferreira (1997) also pointed to the proximity of the terra firme as influencing the increase in richness and diversity in the igapó and Wittmann *et al.* (2002) attributed the high richness recorded in high várzea forests also to the contribution of species shared with the terra firme.

Conversely, since P8 had the highest average flood height and duration and the lowest range of flood height variation (0.6 m), it consequently had the lowest richness and diversity among the study plots.

The overall diversity of the set of plots, considering Fisher's coefficient (Fisher's α), was higher than that found by Targhetta (2012) in the Uatuma River igapó (13.59). Considering the Shannon index, the diversity was higher than that found by Hamaguchi (2009) in the igapó of the Tupé RDS (2.95).

Table 6 - Comparison of the richness and diversity obtained by different authors in various locations and ecosystems throughout the Amazon basin (DBH ≥ 10 cm). Missing data was not presented by the authors.

Author	Ecosystem	Location	Total area (ha)	Total wealth	Maximum wealth ha^{-1}	Total diversity	
						Fisher	Shannon
This study	Igapó	Jaù River	10	193	74	39,34	4,19
Ferreira (1997)	Igapó	Jaù River	3	-	137	-	-
Hamaguchi (2009)	Igapó	Lake Tupé	1,01	53	-	-	2,95
Ferreira & Parolin (2011)	Igapó	Taruma-Mirim River	1,44	102	-	-	-
Montero *et al.* (2014)	Igapó	Rio Negro	10	243	79	-	-
Targhetta (2012)	Igapó	Uatuma River	3	69	62	13,59	-
Wittmann *et al.* (2002)	Và rzea	RDS Mamirauà	3	226	177	-	-
Wittmann *et al.* (2004)	Và rzea	RDS Mamirauà	5,24	222	172	-	
Balslev *et al.* (1987)	Và rzea	Napo River	1	149	149	83	-
Nebel *et al.* (2001)	Và rzea	Ucayali River	9	321	141	-	-
Luize *et al.* (2015)	Và rzea	Purus River	7,18	328	-	-	-
Stroop *et al.* (2011)	Campinarana	Upper Rio Negro	4	290	116	-	-
Targhetta (2012)	Campinarana	RDS Uatuma	3	122	-	29,33	-
Machado (2009)	Dry land	RDS Amana	3	157	101	-	3,6
Silva *et al.* (2014)	Dry land	Southwest of Amapà	1	125	125	-	4,27
Lima Filho *et al.* (2001)	Dry land	Urucu River Basin	3	577	322	-	-

5.3. Phytosociological characterization

Pouteria elegans was among the three most abundant species both in the present study (most abundant species) and in the study by Ferreira (1997), also carried out on the Jaù River, where the author established 3 plots in three habitats assumed to be different (lake, riverbank and igarapé); and in that by Hamaguchi (2009), carried out in the Tupé RDS. In Ferreira's study, specifically, it was the most abundant in plot 2 (riverbank) and the third most abundant in plot 3 (igarapé). In Hamaguchi's study it was the third most abundant. *Amanoa oblongifolia*, which was the second most abundant in the present study, was the most abundant in plot 1 (lake) and the second most abundant in plot 2 (riverbank) in Ferreira's study. *Macrolobium acaciifolium* was the third most abundant in the present study and the second most abundant in plot 1 (lake) in the Ferreira study.

Ter Steege *et al.* (2013) developed a study in which they classified Amazonian species with an estimated abundance greater than 3.7×10^{8} individuals for the entire biome as hyperdominant (not to be confused with *dominant* in the phytosociological sense). Excluding those of dubious identity and *singletons*, 22 species found in the present study (11.4%) are among those considered hyperdominant by that author (Table 7).

Pouteria elegans was the species with the highest relative density in this study and the one with the third highest relative density in Hamaguchi's study. The same species was the second most frequent in this study and the third most frequent in Hamaguchi's study.

Macrolobium acaciifolium was the second most dominant species both in this study and in Ferreira's study, specifically in plot 1 (lake) in the case of this author's work. *Swartzia polyphylla*, in the present study, was the third most dominant; while in Ferreira's study it was the second most dominant, both in plot 2 (riverbank) and plot 3 (igarapé). The fourth most dominant species in the present study (*Aldina latifolia*) was the most dominant in the study by Parolin *et al.* (2004b), carried out on the Taruma-Mirim river; and in the lower section of the Negro river, in the Anavilhanas region, in the study by Montero *et al.* (2014). In the latter study, however, the species seems to have been identified as *Aldina heterophylla*, a typical grassland species (Anderson *et al.*, 1975; Targhetta, 2012).

Table 7 - Hyperdominant species found, *sensu* Ter Steege *et al.* (2013), and their respective position in the ranking.

Species	***Rankin* of hyperdominance**
Hevea brasiliensis	14ª
Licania heteromorpha	19ª
Micropholis guyanensis	26ª
Licania apetala	27ª
Brosimum guianense	45ª
Goupia glabra	52ª
Virola elongata	55ª
Attalea maripa	59ª
Licania micrantha	62ª
Dialium guianense	67ª
Pouteria elegans	82ª
Handroanthus barbatus	90ª
Sacoglottis guianensis	130ª
Calophyllum brasiliense	133ª
Licania octandra	144ª
Caryocar glabrum	146ª
Couepia paraensis	153ª
Diplotropis purpurea	175ª
Mabea nitida	178ª
Swartzia polyphylla	202ª
Vatairea guianensis	214ª
Macrolobium acaciifolium	226ª

In terms of Importance Value Index (IVI), *Amanoa oblongifolia* was the third most important in the present study and the most important in the study by Targhetta (2012), assuming that the species was the same one identified by the author as *Amanoa* cf. *guianensis*. Twenty-six species found in this study were cited by Ter Steege *et al.* (2013) as being more important in igapó than in other Amazonian forest ecosystems (Table 8).

Table 8 - Species cited by Ter Steege *et al.* (2013) as being most important (IVI) for igapó in relation to other Amazonian ecosystems.

other Amazonian ecosystems.

Family	**Species**
Annonaceae	*Duguetia uniflora*
Annonaceae	*Xylopia parviflora*
Apocynaceae	*Aspidosperma nitidum*
Calophyllaceae	*Caraipa grandifolia*
Chrysobalanaceae	*Licania apetala*
Chrysobalanaceae	*Licania heteromorpha*
Chrysobalanaceae	*Licania micrantha*
Euphorbiaceae	*Alchornea discolor*
Euphorbiaceae	*Hevea spruceana*
Fabaceae	*Aldina latifolia*
Fabaceae	*Clathrotropis nitida*
Fabaceae	*Crudia amazonica*
Fabaceae	*Heterostemon mimosoides*
Fabaceae	*Hydrochorea marginata*
Fabaceae	*Macrolobium acaciifolium*

Fabaceae	*Macrolobium multijugum*
Fabaceae	*Parkia discolor*
Humiriaceae	*Humiriastrum cuspidatum*
Malpighiaceae	*Byrsonima incarnata*
Meliaceae	*Trichilia rubra*
Myrtaceae	*Eugenia gomesiana*
Salicaceae	*Laetia suaveolens*
Salicaceae	*Homalium racemosum*
Sapotaceae	*Elaeoluma glabrescens*
Sapotaceae	*Elaeoluma nuda*
Sapotaceae	*Pouteria elegans*

With regard to the most abundant families, the results found were similar to those obtained by Ferreira (1997), with the Fabaceae and Sapotaceae families being, respectively, the most abundant and second most abundant for plots 2 (riverbank) and 3 (igarapé) in that author's study. In plot 1 (lake), in Ferreira's study, Euphorbiaceae appears as the most abundant, but it is believed that the author included in this quantity the Phyllanthaceae family, predominantly represented by the *Amanoa oblongifolia* species in the Jaù river igapó. By separating Euphorbiaceae from Phyllanthaceae, it is estimated that Fabaceae also

had been the most abundant family found by the author in that plot. However, the Phyllanthaceae family alone stood out as the third most abundant in this study. In the study by Hamaguchi (2009), Fabaceae was also the most abundant family.

Similar to Hamaguchi's study, Fabaceae was also the richest family in this study. The family has been reported as having the highest richness and abundance in other floristic studies carried out in the Amazon as a whole (Gentry, 1988; Ter Steege *et al*., 2013). The family was also the most dominant and important in the present study and in others carried out in igapó, such as those by Ferreira (1997), Hamaguchi (2009) and Montero *et al*. (2014). Studies carried out in the vàrzea (Wittmann *et al*., 2006) and in the paleo-vàrzea (Assis *et al*., 2015) have also reported the family as the most important.

Lecythidaceae stood out as the second most dominant family and we attribute this to the contribution of the species *Eschweilera tenuifolia*, especially in P8. As a result of the large number of individuals recorded with trunk reiterations, there is an increase in the cross-sectional area of each individual and, consequently, an increase in the basal area of the species and the relative dominance of the family.

5.4. Influence of hydro-edaphic factors

Ferreira (1997), in his study also carried out at PARNA Jaù, although he did not statistically obtain a correlation between the duration of flooding and richness and diversity, he did observe significant differences between the average values observed in the three types of habitat considered (lake, riverbank and igarapé), with the duration of flooding being inversely proportional to richness and diversity. The same author reports, however, that for the same topographic interval, but considering igapós from different regions, the variation in richness is not always significantly influenced by the duration of the flood (Ferreira, 2000). Although this study, a *priori,* did not consider the existence of distinct habitats in the igapó, the negative influence of flooding on richness and diversity was also found. Targhetta (2012), in the igapó of the Uatuma River, similarly to the present study, also found a significant negative correlation between flood duration and richness.

With regard to community structure, Ferreira (2000) found significant differences between the mean basal area of low, medium and high topographic levels in the Jaù River igapó. However, the variation in basal area seems to be poorly correlated with the variation in flood height in that author's study, since the highest averages were recorded in the middle levels. Targhetta (2012) also found a significant correlation between flood duration and basal area. This study found a significant positive correlation between basal area and flood duration, with an increase in mean values per quadrant as flood duration increased. This pattern is the opposite of that observed in the marshland by Wittmann *et al.* (2004), where sites with a longer duration of flooding had a lower basal area than those that were less flooded. In the present study, this positive correlation between flooding and basal area may be related to competition: on more flooded sites the individuals are under less competitive pressure, since fewer species are adapted to high flooding; as a result, the trees have more space for growth in diameter. The main difference in this case is that the species most adapted to high flooding are mostly pioneers (Parolin *et al.*, 2002). As such, they are species that invest more in longitudinal than transverse growth. In igapó, pioneer species are scarce, since the low geomorphological dynamics (Wittmann *et al.*, 2004), combined with low fertility (Furch & Junk, 1997) and consequent low productivity (Junk *et al.*, 2015), do not promote the establishment of a clear ecological succession under normal circumstances.

In relation to community structure according to the flood gradient, unlike the present study, Targhetta (2012) found no correlation between species composition and the duration of flooding in the Uatuma River igapó. The author draws attention, however, to irregularities identified in the flood pulse of the Uatuma River, as a consequence of the Balbina Hydroelectric Power Plant. This influence has led to a loss of flood seasonality, especially at the lower and upper ends of the topographic gradient, profoundly affecting the dynamics of the igapó biota in that river. Considering individuals with DBH $\geq$ 5 cm, Ferreira (2000) highlighted the existence of species overlap along the flood gradient in a study conducted on the Jaù River and the Taruma-Mirim River, using plots classified *a priori* according to three topographic levels, contemplating *a* range of variation of 5 meters. On the other hand, Ferreira & Parolin (2011) reported a clear change in species composition along a topographic gradient with the same amplitude as the previous study, in the Taruma-Mirim river.

The present study took into account a greater amplitude of flood duration and, consequently, of the corresponding topographic gradient (respectively > 280 days per year^{-1} and 11 meters). In addition, it covered a larger area than the three studies mentioned above, which allowed for a better identification of different occurrences of species along the gradient, as detailed above. Finally, with the exception of the work by Targhetta (2012), data collection was not carried out taking into account the delimitation of plots according to habitats assumed to be distinct or specific topographic intervals. This *a priori* classification is not objectively justified.

Among the species restricted to higher elevations (low tolerance to flooding), 24 were cited by Ribeiro *et al.* (1999) as also occurring in terra firme forests. In this group are species that were restricted to floods of less than 1 meter in height, located in sites flooded only in extreme flood events. The contribution of terra firme species at higher elevations was also mentioned by Wittmann *et al.* (2002) for highland forests, but the number of species shared in the latter case was higher, taking into account the work of Ribeiro *et al.* (1999) (46 species).

Among the species classified as being of intermediate tolerance to flooding, *Vatairea guianensis* was also cited by Wittmann *et al.* (2004) as being of intermediate association between the lowland and highland floodplain communities. Although the studies on varzea (e.g.: Ayres, 1993; Wittmann *et al., 2002;* Wittmann *et al.,* 2004) did not describe a well-defined species composition for the mid-elevations of the varzea forest that corresponded to a phytophysiognomy that was particularly distinct from low and high varzea (what could be called mid-elevation), some species in the present study, although not restricted, were clearly more abundant at mid-elevations. These included *Endlicheria levelii*, *Erythroxylum spruceanum* and *Calyptranthes* cf. *spruceanum*. These species, however, were not among the most dominant at these elevations, thus not promoting the constitution of a distinct phytophysiognomy for the middle elevations. On the other hand, it was observed that two species of non-woody palms, although not quantified in this study, were characteristic and almost restricted, respectively, to the intermediate and high elevations: *Leopoldinia pulchra* and *Astrocaryum acaule*. Unlike what was described by Piedade *et al.* (2006) for the Anavilhanas igapó, no dense populations of *Astrocaryum jauari* were observed in the Jaù river. The occurrence of dense stands of this species, however, is associated by the authors with the existence of successional stages, which indicates an exception for the Anavilhanas igapó in relation to igapó in general.

Of the 10 species classified as having high flood tolerance, *Acosmium nitens, Licania apetala* and *Macrolobium acaciifolium* are also typical of lowland rainforests and *Buchenavia ochroprumna* has a congeneric species in this environment (*B. oxycarpa*) (Wittmann *et al.*, 2010b).

Despite the proposed classification of species' tolerance to flooding, based on their greater abundance at certain intervals along the gradient, their distribution along the gradient generally does not allow for the delimitation of zones. This is because the limits of occurrence along the gradient are not well defined. Worbes (1997) and Keel & Prance (1979) had already mentioned the existence of gradual transitions along the gradient in the igapó, without abrupt limits. Junk *et al.* (2015) point out that in igapó there are a large number of species with similar resistance to flooding, with restrictions on the occurrence of species only in conditions of extreme flooding.

One of these exceptional situations was observed in this study. Although the species *Eschweilera tenuifolia was* representative throughout most of the flooding gradient, it tended to be restricted to highly flooded sites, with 70% of the individuals found in sites under flooding of more than 10 m (up to more than 280 days a year^{-1} of flooding). The species was practically dominant at this level (98% of the individuals), which allowed for the clear delimitation of a specific zone in the igapó, called a *macacarecuial in* this work, in reference to the common name of the species (macacarecuia). The existence of these *macacarecuials* partly contradicts the general notion that there is no zonation of species along the flood gradient in the igapó. The presence of sites monodominated by the species in question in the igapó of the Jaù river was recently mentioned by Junk *et al.* (2015).

The high dominance of a few species in areas subject to long periods of flooding is a pattern also described for floodplain forests (Schongart *et al.*, 2000; Parolin *et al.*, 2002; Wittmann & Junk, 2003; Wittmann *et al.,* 2006). The *macacarecuial*, therefore, resembles the early secondary succession forests dominated by *Cecropia latiloba,* due to the fact that both are environments considered to be little disturbed by anthropic action (Parolin *et al.*, 2002), with a little stratified canopy (Worbes *et al.*, 1992; Wittmann *et al.,* 2002) and clearly differentiated from other igapó and vàrzea forest formations respectively. They differ, however, because the *macacarecuial* has a high basal area and low tree density and because *Eschweilera tenuifolia*, unlike *Cecropia latiloba, is* not a pioneer species, which is characterized by rapid growth and a vast number of widely disseminated small seeds (Goulding, 1983; Parolin *et al.,* 2002). *Eschweilera tenuifolia*, in fact, is reported to be very slow-growing, with the age of some individuals estimated at over 1000 years (Schongart, unpublished data).

The *macacarecuial*, in terms of the ecological characteristics of the dominant species, without taking into account structural similarities, resembles, on the other hand, the varzea vegetation called *chavascal*. Both formations consist of slow-growing species and are not part of the successional series (Wittmann *et al.*, 2002; Wittmann *et al.*, 2004). In igapó, in fact, unlike vàrzea in general (*chavascal* is an exception), the process of ecological succession occurs on a very slow time scale, being practically imperceptible or non-existent, due to the relative stability and low productivity of this environment (Sioli, 1975; Junk, 1989; Junk *et al.,* 2015). This is not the case in the uplands, where geomorphological dynamics are very intense and ecological succession occurs on a much shorter time scale, resulting in different types of environments corresponding to different successional stages (Worbes, 1997; Wittmann *et al.*, 2002; Wittmann *et al.*, 2004). Ecological succession and differences in flood height are the two main factors responsible for the distribution and structure of the floodplain community (Junk, 1989; Worbes, 1997; Ayres, 1993).

Different successional stages in the floodplain also correspond to different levels of granulometry in the soil, as the advance of succession promotes a decrease in the size of the sediment particles that are decanted, due to the decrease in the speed of the water current (Wittmann *et al.*, 2004). This results in the formation of a granulometric gradient that varies from sandy at lower elevations to clayey at higher elevations, with an important influence on the zonation of species from the river banks. This pattern of variation in the granulometric gradient showed the opposite trend in this study, with the lower elevations tending to be more clayey than the higher ones. The reason for this opposite trend is unclear. Although Worbes (1986, 1997) observed marked differences in floristic composition between predominantly clayey and sandy soils in igapó, there was little community structure in the present study, with few species restricted to specific granulometric levels.

This study, however, prioritized the analysis of the influence of the edaphic gradient as a whole (intrinsically related physico-chemical variables) (Appendix B). Similar to Targhetta's work (2012), no correlation was found between the edaphic gradient and the diversity and structure of the vegetation in the igapó.

(except for the significant but low correlation with density). In general, however, the edaphic gradient is not very important in structuring the igapó community. The results obtained here and those obtained by Furch (1997) in the Taruma-Mirim River igapó indicate that there is a tendency for igapó soils, considered on a relatively small geographical scale (e.g. the Jaù River basin and the Taruma-Mirim River basin separately), to be relatively homogeneous, regardless of topography. On a continental scale, Montero *et al.* (2014) point to edaphic variations, among other factors, as being responsible for the considerable variation in floristic composition along different sections of the Rio Negro. Wittmann *et al.* (2012) also mentions that the quality of the substrate influences the species composition of the floodplain on a continental scale.

6. Conclusion

The importance of edaphic factors has been shown to be small or almost nil in structuring the igapó tree community, contrary to the predictions of some authors. However, as discussed, this importance tends to increase as larger geographical scales are taken into account.

The influence of flooding, on the other hand, proves to be important in this structuring, being correlated with both diversity and composition and structure. However, only one well-defined and characterized vegetation zone was identified in the Jaù River igapó, and it was restricted to the extreme upper level of the flood gradient. More in-depth studies, taking into account the vertical structure of the forest, wood density values, the age of the stands and the different occurrence of other vegetation components (e.g. non-timber palms, lianas, etc.) could help in the delimitation and characterization of other potential typologies. The predominantly abundant species along certain intervals of the gradient can also be useful in characterizing these typologies.

The results obtained show that a more detailed approach, covering the entire flood gradient, is better able to characterize the variation in species composition in igapó than restricted approaches. Furthermore, the prior qualification of samples, based on topographic intervals and/or unjustified typification of habitats, can lead to inconsistent results in terms of describing the variation in composition according to the duration/height of the flood.

Given the high beta-diversity between igapó forests considered on a larger scale, as demonstrated in recent work, more studies are needed to define whether or not there are similar or equivalent patterns in the variation of floristic composition according to the flooding gradient for igapós in other regions.

7. BIBLIOGRAPHICAL REFERENCES

Adis, J. 1984. Seasonal igapó forests of central Amazonian black-water rivers and their terrestrial arthropod fauna. *In*: Sioli, H. (Ed). *The Amazon - Limnology and Landscape Ecology of a Mighty Tropical River and its Basin*. Dr. W. Junk Publishers, Dordrecht/Boston/Lancaster. p. 245-268.

Anderson, A.B.; Prance, G.T.; Albuquerque, B.W.P. 1975. Studies on the vegetation of Amazonian Campinas III: The woody vegetation of the Campina of the INPA-SUFRAMA Biological Reserve (Manaus-Caracarai, km 62). *Acta Amazonica*, 5: 225246.

APG (Angiosperm Phylogeny Group) III. 2009. An update of the Angiosperm Phylogeny Group classification for the orders and families of flowering plants: APG III. *Bot. J. Linnean Soc.*, 161: 105-121.

Assis, R.L.; Haugaasen, T.; Schongart, J.; Montero, J.C.; Piedade, M.T.F.; Wittmann, F. 2015. Patterns of tree diversity and composition in Amazonian floodplain paleo-vàrzea forest. *Journal of Vegetation Science*, 26: 312-322.

Ayres, J.M. 1993. The varzea forests of Mamirauà. *In* : Mamirauà Civil Society (Ed). *Mamirauà Studies.* Vol. 2. Sociedade Civil Mamirauà, Belém, 1993, p. 1123.

Balslev, H.; Lutteyn, J.; Yllgaard, B.; Holm-Nielsen, L. 1987. Composition and structure of adjacent unflooded and floodplain forest in Amazonian Ecuador. *Opera Botanica*, 92: 37-57.

Bongers, F.; Engelen, D.; Klinge, H. 1985. Phytomass structure of natural plant communities on spodosols in southern Venezuela: the Bana woodland. *Vegetatio*, 63: 13-24.

Campbell, D.G.; Daly, D.C.; Prance, G.T.; Maciel, U.N. 1986. Quantitative ecological inventory of terra firme and vàrzea tropical forest on the Rio Xingu, Brazilian Amazon. *Brittoniana*, 38: 369-393.

Campbell, D.G.; Stone, J.L.; Rosas, J.A. 1992. A comparison of the phytosociology and dynamics of three (vàrzea) forests of known ages, Rio Juruà, Western Brazilian

Amazon. *Journal of Linnean Society*, 108: 213-237.

Colwell, R.K.; Coddington, J.A. 1994. Estimating terrestrial biodiversity through extrapolation. *Phil. Trans. R. Soc. Lond.*, 345: 101-118.

Costanza, R.; d'Arge, R.; de Groot, R.; Farberk, S.; Grasso, M.; Hannon, B.; Limburg, K.; Naeem, S.; O'Neill, R.V.; Paruelo, J.; Raskin, R.G.; Sutton, P.; van den Belt, M. 1997. The value of the world's ecosystem services and natural capital. *Nature*, 387.

Curtis, J.T.; McIntosh, R.P. 1951. An upland forest continuum in the prairie-forest border region of Wisconsin. *Ecology*, 32: 476-496.

De Groot, R.; Brander, L.; Ploeg , S.; Costanza, R.; Bernard, F.; Braat, L.; Christie, M.; Crossman, N.; Ghermandi, A.; Hein, L.; Hussain, S.; Kumar, P.; McVittie, A.; Portela, R.; Rodriguez, L.C.; Brink, P.; vanBeukering, P. 2012. Global estimates of the value of ecosystems and their services in monetary units. *Ecosystem Services*, 1: 50-21.

De Simone, O.; Haase, K.; Müller, E.; Junk, W.J.; Gonsior , G.A.; Schmitt, W. 2002. Impact of root morphology on metabolism and oxygen distribution in roots and rhizosphere from two Central Amazon floodplain tree species. *Funct Plant Biol*, 29:1025-1035.

Ducke, A.; Black, G.A. 1954. Notes on the phytogeography of the Brazilian Amazon. *Bol Técn. Inst. Agron.*, 29: 1-62.

Ferreira, L.V. 1997. Effects of the duration of flooding on species richness and floristic composition in three hectares in the Jaù National Park in floodplain forests in central Amazonia. *Biodiversity and Conservation*, 6: 1353-1363.

Ferreira, L.V.; Prance, G.T. 1998. Structure and species richness of low-diversity floodplain forest on the Rio Tapajós, Eastern Amazonia, Brazil. *Biodiversity and Conservation*, 7: 585-596.

Ferreira, L.V. 2000. Effects of flooding duration on species richness, floristic composition and forest structure in river margin habitat in Amazonian blackwater floodplain forests: implications for future design of protected areas. *Biodiversity and Conservation*, 9: 1-14.

Ferreira, L.V.; Parolin, P. 2011. Effects of flooding duration on plant demography in a black-water floodplain forest in Central Amazonia. *Pesquisa. Botânica*, 62: 323332

Ferreira, C.S.; Piedade, M.T.F.; Franco, A.C.; Gonçalves, J.F.C.; Junk, W. 2009. Adaptive strategies to tolerate prolonged flooding in seedlings of floodplain and upland populations of Himatanthus sucuuba, a Central Amazon tree. *Aquatic Botany*, 90: 246-252.

Fisher, R.A.; Corbet, A.S.; Wiliams, C.B. 1943. The relation between the number of species and the number of individuals in a random sample of an animal population. *Journal of Animal Ecology*, 12: 42-58.

Vitória Amazônica Foundation. 2008. Jaù National Park Management Plan. Fundaçao Vitória Amazônica/Instituto Brasileiro do Meio Ambiente e dos Recursos Naturais Renovâveis, Manaus.

Furch, K. 1997. Chemistry of vàrzea and igapó soils and nutrient inventory in their floodplainforests. *In*: *The Central Amazon floodplain: Ecology of a pulsing system*. Junk, W.J. (Ed). Ecolog. Stud. v. 126. Springer, Berlin/Heidelberg/ New York, p. 4768.

Furch, K.; Junk, W.J. 1997. The chemical composition, food value and decomposition of herbaceous plants and leaf-litter of the floodplain forest. *In*: Junk, W.J. (Ed). *The Central Amazon floodplain: ecology of a pulsing system*. Ecolog. Stud. v. 126. Springer, BerlinHeidelberg/ New York, p. 187-206.

Gentry, A.H. 1988. Changes in plant community diversity and floristic composition on environmental and geographical gradients. *Ann. Mo. Bot. Gard.*, 75: 1-34.

Godoy, J.R.; Petts, G.; Salo, J. 1999. Riparian flooded forests of the Orinoco and Amazon basin: a comparative review. *Biodiversity Conservation*, 8: 551-586.

Goulding, M. 1983. The role of fishes in seed dispersal and plant distribution in Amazonian floodplain ecosystems. *Sonderbd. Naturwiss. Hamburg*, 7: 271-283.

Hamaguchi, J.O. 2009. *Structure and floristic composition of tree and shrub species in an igapó forest at Lago Tupé, Manaus, AM.* Master's dissertation, National Institute for Amazonian Research/Federal University of Amazonas, Manaus, Amazonas. 85 pp.

Hammer, 0.; Harper, D.A.T.; Ryan, P. D. 2001. PAST: Paleontological Statistics Software Package for Education and Data Analysis. *Palaeontologia Electronica,* 4(1): 9p.

Haugaasen, T.; Peres, C.A. 2006. Floristic, edaphic, and structural characteristics of flooded and unflooded forests in the lower Rio Purus region of central Amazonia, Brazil. *Acta Amazonica*, 36(1): 25-36.

HidroWeb (Hydrological Information System), 2015 (hidroweb.ana.gov.br/). Accessed: 15/05/15.

Ilhardt, B.L.; Verry, E.S.; Palik, B.J. 2000. Defining riparian areas. *In*: Verry, E.S.; Hornbeck, J.W.; Dolloff, C.A. (Ed). *Riparian management in forests of the continental eastern United States*. Lewis Publishers, Boca Raton/London/New York/Washington DC, p. 23-42.

Inuma, J.J. 2006. *Comparison of the diversity and structure of woody plant communities on terra firme, vàrzea and igapó in Amazonas, Central Amazonia.* Doctoral thesis, National Institute for Amazonian Research, Manaus.

Irion, G. 1976. Quaternary sediments of the upper Amazon lowlands of Brazil. *Biogeographica,* 7: 163-167.

Irion, G. 1978. Soil infertility in the Amazonian rain forest. *Naturwissenschaften*, 65: 515-519.

Irion, G.; de Mello, J.A.S.N.; Morais, J.; Piedade, M.T.F.; Junk, W.J.; Garming, L. 2010. Development of the Amazon valley during the Middle to Late Quaternary: sedimentological and climatological observations. *In*: Junk, W.J.; Piedade, M.T.F.; Wittmann, F.; Schongart, J.; Parolin, P. (Eds). *Amazonian Floodplain Forests - Ecophysiology, Biodiversity and Sustainable Management*. Springer,

Berlin/Heidelberg/New York, p. 27-42.

Junk, W.J. 1980. Floodplains - A Challenge for Limnology. *Acta Amazonica,* 10(4): 775-795.

Junk, W.J. 1989. Flood tolerance and tree distribution in central Amazonian floodplains. *In*: Holm- Nielsen, L.B.; Nielsen, I.C.; Balslev, H. (Eds). *Tropical forests: botanical dynamics, speciation and diversity*. Academic Press, New York, p. 47-64.

Junk, W.J.; Barley, P.B.; Sparks, R.E. 1989. The flood-pulse concept in riverfloodplain systems. *Can Spec. Publ. Fish Aquat. Sci.*, 106: 110-127.

Junk, W.J. 1993. Wetlands of tropical South America. *In*: Whigham, D.; Hejny, S.; Dykyjova, D. (Eds). *Wetlands of the world.* Junk Publications, Dordrecht, p. 679739.

Junk, W.J.; Piedade, M.T.F. 1997. Plant life in the floodplain with special reference to herbaceous plants. *In*: Junk, W.J. (Ed). *The central Amazon floodplain: ecology of a pulsating system*. Ecolog. Stud. v. 126. Springer, Berlin/Heidelberg/New York. p. 147-186.

Junk, W.J.; Piedade, M.T.F. 2010. An Introduction to South American Wetland Forests: Distribution, Definitions and General Characterization. *In*: Junk, W.J.; Piedade, M.T.F.; Wittmann, F.; Schongart, J.; Parolin, P. (Eds). *Amazonian Floodplain Forests - Ecophysiology, Biodiversity and Sustainable Management*. Springer, Berlin/Heidelberg/New York, p.3-26.

Junk, W.J.; Piedade, M.T.F.; Schongart, J.; Cohn-Haft, M.; Adeney, J.M.; Wittmann, F. 2011. A classification of major naturally-occurring Amazonian lowland wetlands. *Wetlands*, 31: 623-640.

Junk, W.J.; Piedade; M.T.F.; Lourival, R.; Wittmann, F.; Kandus, P.; Lacerda, L.D.; Bozelli, R.L.; Esteves, F.A.; Nunes da Cunha, C.; Maltchik, L.; Schongart, J.; Schaeffer-Novelli, Y.; Agostinho, A.A. 2014. Brazilian wetlands: their definition, delineation, and classification for research, sustainable management, and protection. *Aquatic Conserv: Mar. Freshw. Ecosyst.*, 24: 5-22.

Junk, W.J.; Wittmann, F.; Schongart, J.; Piedade, M.T.F. 2015. A classification of the major habitats of Amazonian black-water river floodplains and a comparison with their white-water counterparts. *Wetlands Ecol. Manage.*, 23.

Keel, S.H.K.; Prance, G.T. 1979. Studies of the vegetation of a white-sand blackwater igapó (Rio Negro, Brazil). *Acta Am* azonica, 9: 645-655.

Klinge, H.; Medina, E. 1979. Rio Negro caatingas and campinas, Amazonas States of Venezuela and Brazil. *In*: Specht, R.L. (Ed). *Heathlands and related shrublands. - Descriptive studies*. Ecosystems of the World. v. 9. Elsevier, Amsterdam, p. 483-488.

Klinge, H.; Junk, W.J.; Revilla, C.J. 1990. Status and distribution of forested wetlands in tropical South America. *Forest Ecol. Manage*., 33/34: 81-101.

Lima Filho, D.A.; Matos, F.D.A.; Amaral, I.L.; Revilla, J.; Coêlho, L.S.; Ramos, J.F.; Santos, J.L. Floristic inventory of terra firme dense ombrophilous forest in the Urucu River region, Amazonas, Brazil. 2001. *Acta Amazonica*, 31(4): 565-579.

Luize, B.G.; Silva, T.S.F.; Wittmann, F.; Assis, R.L.; Venticinque, E.M. Effects of the Flooding Gradient on Tree Community Diversity in Vàrzea Forests of the Purus River, Central Amazon, Brazil. 2015. *Biotropica*, 0 (0): 1-6.

Machado, K.S.S. 2009. *Structure and floristic composition of a terra firme forest in the Amană Sustainable Development Reserve, Central Amazonia.* Master's dissertation, National Institute for Amazonian Research, Manaus, Amazonas. 75 pp.

McCormick, J.F. 1979. A summary of the national riparian symposium. In: U.S. department of agriculture *management of floodplain wetlands and other riparian ecosystems*. Ge Tech Rep WO-, forest service (Ed). *Strategies for protection and* 12 US. Department of Agriculture, Forest Service, Washington DC, p. 362-363.

Melack, J.M.; Hess, L.L. 2010. Remote sensing of the distribution and extent of wetlands in the Amazon basin. *In*: Junk, W.J.; Piedade, M.T.F.; Wittmann, F.; Schongart, J.; Parolin, P. (Eds). *Central Amazonian floodplain forest: ecophysiology, biodiversity and sustainable management*. v. 210. Springer, New York, p. 43-60.

Mitsch, W.J.; Gosselink, J.G. 2000. *Wetlands*. Wiley, New York.

Montero, J.C.; Piedade, M.T.F.; Wittmann, F. 2014. Floristic variation across 600 km of inundation forests (Igapó) along the Negro River, Central Amazonia. *Hydrobiologia*, 729: 229-246.

Moreira, E. 1970. *The Igapós and their use.* University Press, Belém, 1970.

Nebel, G.; Kvist, L.; Vanclay, J.K.; Christensen, H.; Freitas, L.; Ruiz, J. 2001. Structure and floristic composition of floodplain forests in the Peruvian Amazon I. Overstorey. *Forest Ecology and Management*, 150: 27-57.

Novotny, V.; Basset, Y. 2000. Rare species in communities of tropical insect herbivores: pondering the mystery of singletons. *Oikos*, 89: 564-572.

Odum, E.P. 1981. Foreword. *In*: Clark, J.R.; Benforado, J. (Eds). *Wetlands of bottomland hardwood forests*. Elsevier, Amsterdam, p. 8-10.

Oliveira, A.A.; Mori, S.A. 1999. A Central Amazonian terra firme forest - High tree species richness on poor soils. *Biodiv. Conserv*, 8: 1219-1244.

Oliveira, A.N.; Amaral, I.D. 2004. Floristics and phytosociology of a slope forest in Central Amazonia, Amazonas, Brazil. *Acta Amazonica*, 34: 21-34.

Oliveira Wittmann, A.; Piedade, M.T.F.; Parolin, P.; Wittmann, F. 2007. Germination in four lowvarzea tree species of central Amazonia. *Aquat. Bot.*, 86: 197-203.

Oliveira Wittmann, A.; Lopes, A.; Conserva, A.S.; Piedade, M.T.F. 2010. Germination and seedling establishment in floodplain forests. *In*: Junk, W.J.; Piedade, M.T.F.; Wittmann, F.; Schongart, J.; Parolin, P. (Eds). *Central Amazonian floodplain forests: ecophysiology, biodiversity and sustainable management*. Springer, Berlin/Heidelberg/New York.

Oksanen, J.F., Blanchet, G., Kindt, R., Legendre, P., Minchin, P.R., O'Hara, R.B., Simpson, G.L., Solymos, P., Stevens, M.H.H., Wagner, H., 2013. vegan: Community Ecology Package, R package version 2, 0-7 (CRAN.R-project.org/package=vegan).

Parolin, P.; Worbes, M. 2000. Wood density of trees in black water floodplains of Rio Jaù National Park, Amazonia. *Acta Amazonica*, 30(3): 441-448.

Parolin, P.; Oliveira, A.C.; Piedade, M.T.F.; Wittmann, F.; Junk, W.J. 2002. Pioneer trees in Amazonian floodplains: key species form monospecific stands in different habitats. *Folia Geobotan.*, 37: 225-238.

Parolin, P.; De Simone, O.; Haase, K.; Waldhoff, D.; Rottenberger, S.; Kuhn, U.; Kesselmeier, J.; Schmidt, W.; Piedade, M.T.F.; Junk, W. 2004a. Central Amazon floodplain forests: tree survival in a pulsing system. *Botan. Rev.*, 70: 357-380.

Parolin, P.; Adis, J.; Rodrigues, W.A.; Amaral, I.; Piedade, M.T.F. 2004b. Floristic study in na igapó floodplain Forest in Central Amazônia, Brazil (Taruma-Mirim, Rio Negro). *Amazoniana*, XVIII (1/2): 29-47.

Parolin, P. 2009. Submerged in darkness: adaptations to prolonged submergence by woody species of the Amazonian floodplains. *Ann. Bot.*, 103: 359-376.

Piedade, M.T.F. 1985. *Ecology and reproductive biology of Astrocaryum jauari Mart. (Palmae) as an example of a population adapted to the flooded areas of the Negro River.* Master's dissertation, National Institute for Amazonian Research, Manaus.

Piedade, M.T.F.; Parolin. P.; Junk, W.J. 2006. Phenology, fruit production and seed dispersal of Astrocaryum jauari (Arecaceae) in Amazonian black-water floodplains. *Rev Biol Trop*, 54:1171-1178.

Pires, J. M.; Prance, G. T. 1985. The vegetation types of the Brazilian Amazon. *In*: Prance, G. T.; Lovejoy, T. (Ed). *Key environments: Amazonia.* Pergamon Press, p. 109-145.

Prance, G.T. 1979. Notes on the vegetation of Amazonia. v. 3. The terminology of Amazonian Forest types subject to inundation. *Brittonia*, 31: 26-38.

R Development Core Team. 2011. R: A language and environment for statistical computing. R Foundation for Statistical Computing, Vienna, Austria. ISBN 3900051-07-0, URL (www.R-project.org/).

Radambrasil. 1978. Sheet SA.20/Manaus. Natural Resources Survey, v.18. National Department of Mineral Production, Ministry of Mines and Energy, Rio de Janeiro.

Rankin-de-Mérona, J.M.; Prance, G.T.; Hutchings, R.W.; Silva, M.F.; Rodrigues, W.A.; Uehling, M.E. 1992. Preliminary results of a large-scale tree inventory of upland rain forest in the Central Amazon. *Acta Amazonica*, 22: 493-534.

Resende, M.; Curi, N.; Rezende, S. B.; Corrêa, G. F. 2007. *Pedologia - Base para a distinçào de ambientes*. 5. ed. rev. Editora UFLA, Lavras, 2007.

Revilla, J.D.C. 1981. *Floristic and Phytosociological Aspects of Floodplain Forest (igapó). Praia Grande, Rio Negro, Amazonas.* Master's dissertation, National Institute for Amazonian Research, Manaus.

Ribeiro, J.E.; Hopkins, M.; Vicentini, A.; Sothers, C.A.; Costa, M.A.; Brito, J.M.; Souza, M.A.; Martins, L.H.; Lohmann, L.G.; Assunçao, P.A.; Pereira, E.; Silva, C.F.; Mesquita, M.R.; Procópio, L.C. 1999. *Flora da Reserva Ducke - Guide to the identification of vascular plants of a terra firme forest in Central Amazonia*. INPA-DFID, Manaus, 1999.

Richards, P.W. 1952. *The tropical rain forest*. Cambridge University Press, London, 1952.

Ricklefs, R.E. 1990. *Ecology*. WH Freeman, New York, 1990.

Rodrigues, W.A. 1961. *Preliminary study of the high várzea forest of an island in the lower Rio Negro with clayey and humid soil*. Publication No. 10 of the National Institute for Amazonian Research, Manaus.

Rollet, B. 1978. *Architecture and growth of tropical forests*. SUDAM, Belém.

Rosales, J.; Petts, G.; Knab-Vispo, C. 2001. Ecological gradients in riparian forests of the lower Caura River, Venezuela. *Plant Ecology*, 152(1): 101-118.

Rosales, J.; Vispo, C.; Dezzeo, N.; Blanco-Belmonte, L.; Knap-Vispo; Gonzalez, N.; Daza, F.; Bradley, C.; Gilvear, D.; Escalante, G.; Chacón, N.; Petts, G. 2002. Ecohydrology of riparian forests in the Orinoco River basin. *In*: *The ecohydrology of South-American rivers and wetlands*. IAHS Special Publication, n. 6, p. 93-110.

Schlüter, U.B.; Furch, B.; Joly, C.A. 1993. Physiological and anatomical adaptations by young *Astrocaryum jauari* Mart. (Arecaceae) in periodically inundated biotopes of central Amazonia. *Biotropica,* 25: 384-396.

Schongart, J.; Piedade, M.T.F.; Worbes, M. 2000. Successional differentiation in structure, floristic composition and wood increment of whitewater floodplain forests in Central Amazonia. *German-Brazilian Workshop on Neotropical Ecosystems - Achievements and Prospects of Cooperative Research*, Hamburg, p. 589-606.

Schongart, J.; Piedade, M.T.F; Ludwigshausen, S.; Hornas, V.; Worbes, M. 2002. Phenology and stem-growth periodicity of tree species in Amazonian floodplain forests. *J. Trop. Ecol.*, 18: 581-597.

Schongart, J. 2003. DendrochronologischeUntersuchungenin

Uberschwemmungswaldem der vàrzea Zentralamazoniens. *Gottinger Beitrage zur Land- und Forstwirtschaft in den Tropen und Subtropen,* 149, Erich Goltze Verlag, Gottingen.

Schongart, J.; Piedade, M.T.F.; Wittmann, F.; Junk, W.J.; Worbes, M. 2005. Wood growth patterns of *Macrolobium acaciifolium* (Benth.) Benth. (Fabaceae) in Amazonian black-water and white-water floodplain forests. *Oecologia*, 145: 454461.

Schubart, H.O.R. 1977. *Report on the excursion to the Jaù River with a view to setting up a Nature Reserve*. National Institute for Amazonian Research, Manaus, 7 p.

Schubart, H.O.R. 1983. Ecology and use of forests. *In*: Prance, G.T. & Lovejoy, T.E. (Eds). *Amazonia*. Editora Brasiliense, Sao Paulo, p. 101-143.

Scolforo, J.R.S.; Pulz, F.A.; Melo, J.M. 1998. Modeling production, age of native forests, spatial distribution of species and structural analysis. *In* : Scolforo, J.R.S. (Org.). *Manejo Florestal*. UFLA/FAEPE, Lavras, p. 189-246.

Shepherd, G.J. 2010. *Fitopac - User manual*. Department of Botany, State University of Campinas, Campinas.

Silva, W.A.S.; Carim, M.J.V.; Guimaraes, J.R.S.; Toste, L.C.L. 2014. Floristic composition and diversity in a stretch of terra firme forest in southwestern Amapà State, Eastern Amazonia, Brazil. *Biota Amazônia*, 3 (4): 31-36.

Sioli, H. 1956. Über Natur und Mensch im brasilianischen Amazonasgebiet. *Erdkunde*, 10(2): 89-109.

Sioli, H. 1975. Tropical river: The Amazon. *In*: Whitton, B. A. (Ed). *River Ecology*. California, California, USA, p. 461-488.

Stropp, J.; Van Der Sleen, P.; Assunçao, P.A.; Silva, A.L.; Ter Steege, H. 2011. Tree communities of white-sand and terra-firme forests of the upper Rio Negro. *Acta Amazonica*, 41: 521-544.

Takeuchi, M. 1962. The structure of the Amazonian vegetation. v. 6. Igapó. *J Fac Sci Univ Tokyo Sect. Bot.*, 3: 297-304.

Targhetta, N. 2012. *Floristic and structural comparison between igapó and campinarana forests along hydro-edaphic gradients in the Calumò Sustainable Development Reserve, Central Amazonia.* Master's dissertation, National Institute for Amazonian Research, Manaus, Amazonas. 105 pp.

Taylor, L.R. 1978. Bates, Williams, Hutchinson - a variety of diversities. *In* : Mound, L.A; Warloff, N. (Eds). *Diversity of insect faunas*. Blackwell, Oxford, 1978, p. 1-18.

Ter Steege, H.; Pitman, N.; Phillips, O.L.; Chave, J.; Sabatier, D.; Duque, A.; Molino, J.F.; Prévost, M.F.; Spichiger, R.; Castellanos, H.; Hildebrand, P.; Vàsquez, R. 2006. Continental scale patterns of canopy tree composition and function across Amazonia. *Nature*, 443.

Ter Steege, H ..... [*et al.*]. 2013. Hyperdominance in the Amazonian Tree Flora.

Nature, 342.

Veloso, H.P.; Filho, A.L.R.R.; Lima, J.C.A. 1991. *Classification of Brazilian vegetation, adapted to a universal system.* Rio de Janeiro, IBGE, Department of Natural Resources and Environmental Studies. 124pp.

Vieira, L. S.; Santos, P. C. T. C. 1987. *Amazonia - Its soils and other natural resources*. Editora Agronòmica Ceres, Sao Paulo, 1987.

Waldhoff, D.; Furch, B; Junk, W.J. 2002. Fluorescence parameters, chlorophyll concentration, and anatomical features as indicators for flood adaptation of an abundant tree species in central Amazonia: *Symmeria paniculata. Environ. Exp. Bot.*, 48: 225-235.

Wittmann, F. 2001. *Artenverbreitung und Bestandesstruktur in amazonischen Vàrzea-Waldern und Moglichkeiten der Erfassung von Waldtypen mittels fernerkundlichen Methoden*. Doctoral thesis, Universitat Mannheim, Fachbereich Geographie.

Wittmann, F.; Anhuf, D.; Junk,W.J. 2002. Tree species distribution and community structure of central Amazonian vàrzea forests by remote-sensing techniques. *J Trop Ecol,* 18: 805-820.

Wittmann, F.; Junk, W.J. 2003. Sapling communities in Amazonian white-water forests. *J. Biogeogr.*, 30: 1533-1544.

Wittmann, F.; Junk, W.J.; Piedade, M.T.F. 2004. The vàrzea forests in Amazonia: flooding and the highly dynamic geomorphology interact with natural forest succession. *Forest Ecology Management*, 196: 199-212.

Wittmann, F.; Schongart, J.; Montero, J.C.; Motzer, M.; Junk, W.J.; Piedade, M.T.F.; Queiroz, H.L.; Worbes, M. 2006. Tree species composition and diversity gradients in white-water forests across the Amazon Basin. *J. Biogeogr.,* 33: 1334-1347.

Wittmann, F.; Schongart, J.; Junk, W.J. 2010a. Phytogeography, species diversity, community structure and dynamics of central Amazonian floodplain forests. *In*: Junk, W.J.; Piedade, M.T.F.; Wittmann, F.; Schongart, J.; Parolin, P. (Ed). *Central Amazonian floodplain forests: ecophysiology, biodiversity and sustainable management*. Springer, Berlin/Heidelberg/New York.

Wittmann, F.; Schongart, J.; Brito, J.M.; Oliveira Wittmann, A.; Parolin, P.; Piedade, M.T.F.; Junk, W.J.; Guillaumet, J.L. 2010b. *Manual de árvores de vàrzea da Amazônia Central - Taxonomy, ecology and use.* INPA Publishing House, Manaus,

2010.

Wittmann, F.; Householder, E.; Piedade, M.T.F.; Assis, R.L.; Schongart, J.; Parolin, P.; Junk, W.J. 2012. Habitat specifity, endemism and the neotropical distribution of Amazonian white-water floodplain trees. *Ecography*, 35: 1-18.

Worbes, M. 1986. Lebensbedingungen und Holzwachstum in zentralamazonischen Uberschwemmungswaldem. Erich Goltze, Gottingen. *Scripta Geobotanica, 17:* 1112.

Worbes, M. 1989. Growth rings, increment and age of tree in inundation forest, savannas and a mountain forest in the Neotropics. *IAWA Bull*, 10(2):109-122.

Worbes, M.; Klinge, H.; Revilla, J.D.; Martius, C. 1992. On the dynamics, floristic subdivision and geographical distribution of vàrzea forests in Central Amazonia. *Journal of Vegetation Science*, 3: 553-564.

Worbes, M. 1997. The forest ecosystem of the floodplains. *In*: Junk, W.J. (Ed). *The central Amazon floodplain: ecology of a pulsating system*. Ecolog. Stud. v. 126. Springer, Berlin/ Heidelberg/New York, p. 223-265.

8. APPENDICES

Appendix A - Relative densities according to diametric classes in the plots.

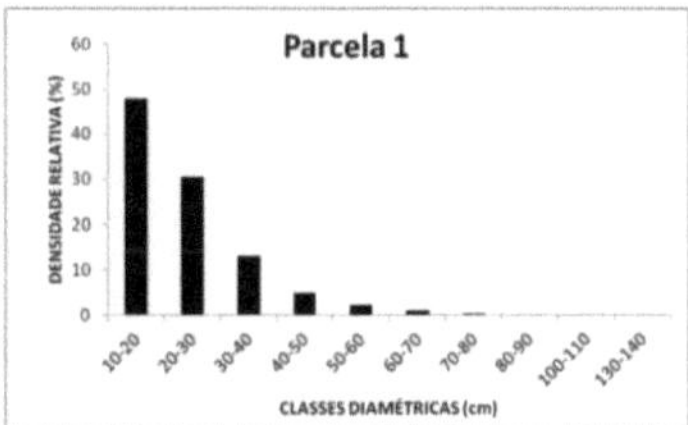

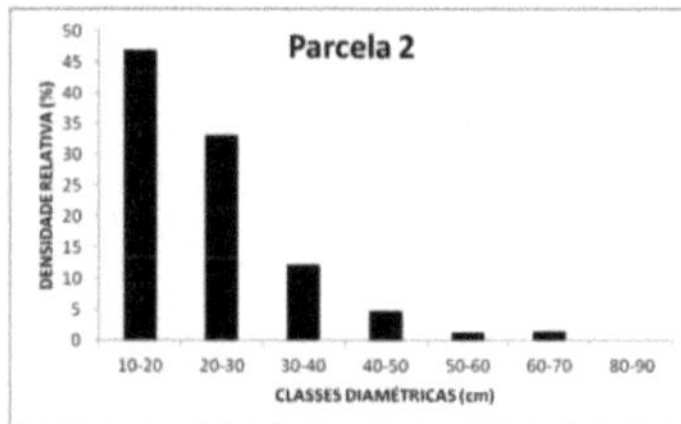

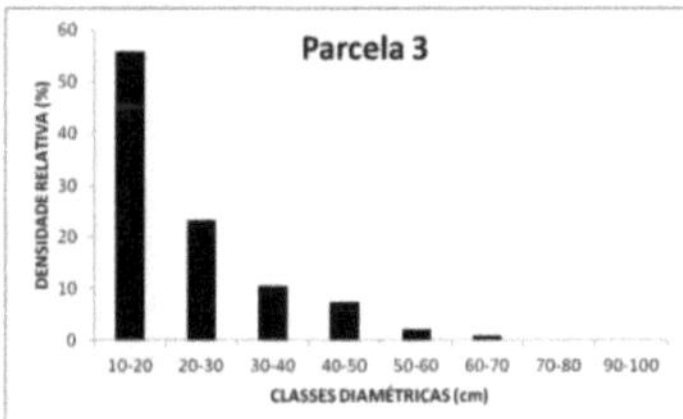

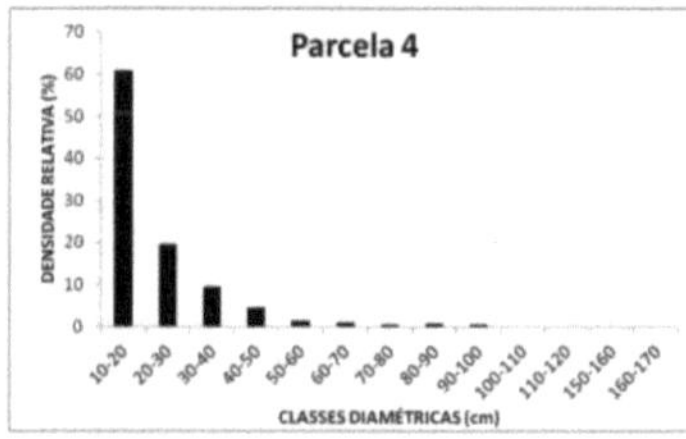

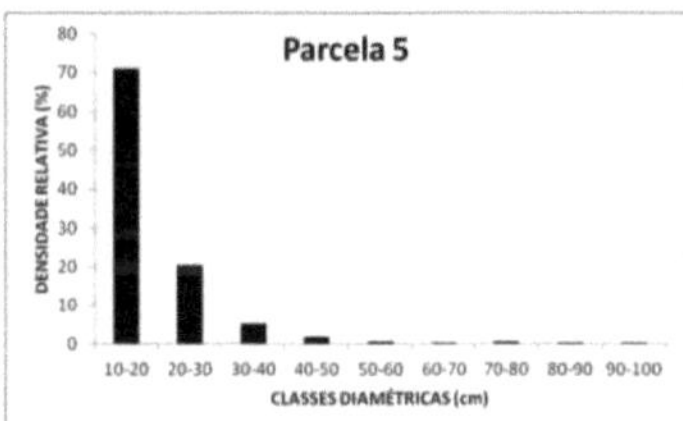

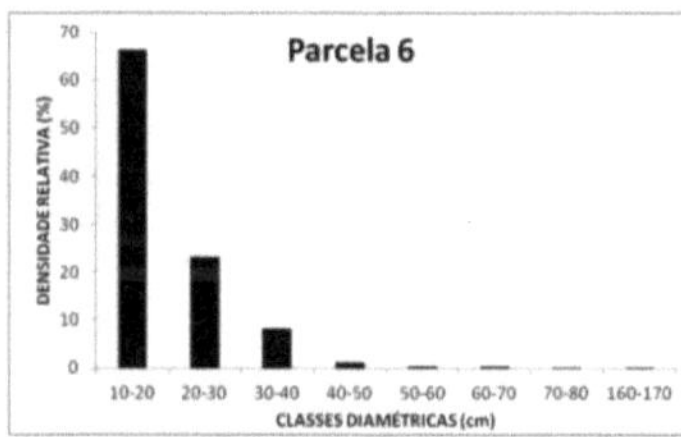

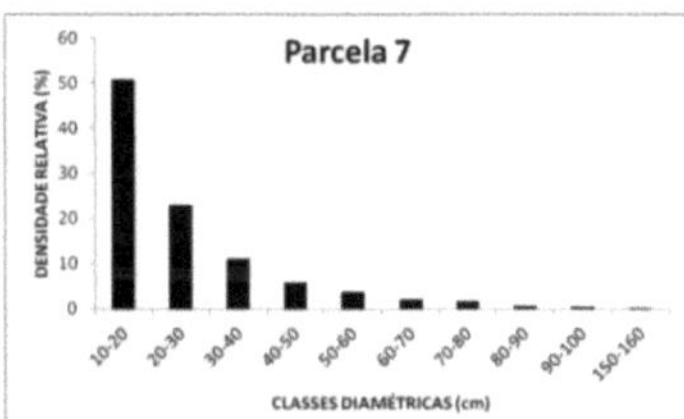

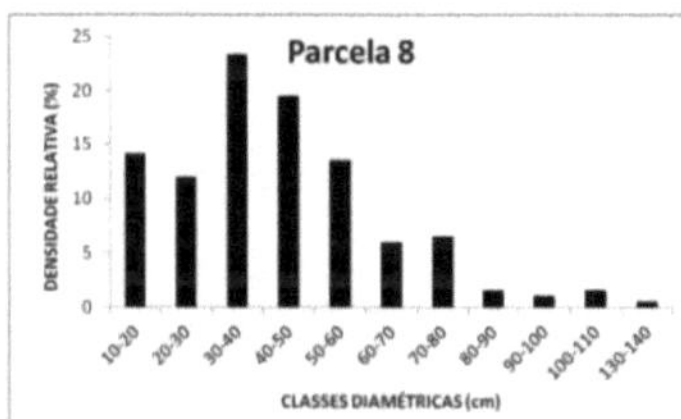

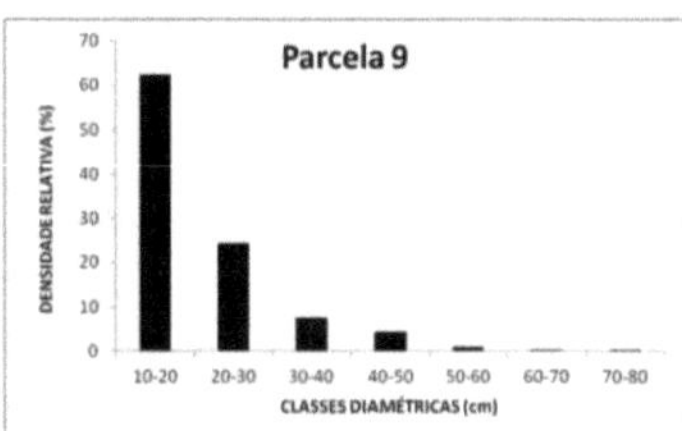

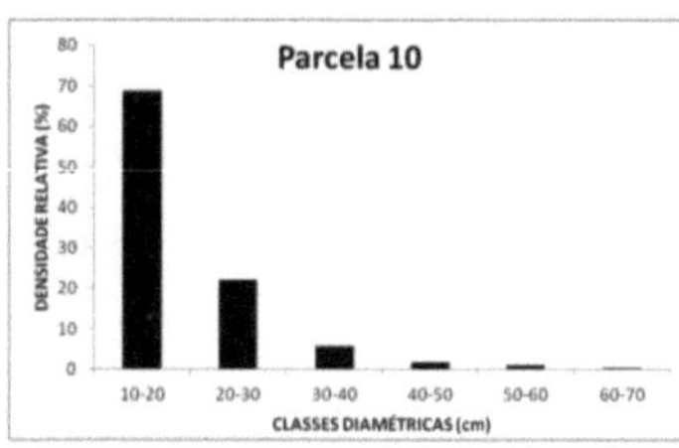

Appendix B - List of families and species recorded and their phytosociological parameter values.

Family/Species	Abund.	Relative Density Relative	Relative Frequency Relative	Dom. Relative	IVI
Annonaceae	**160**	**2,29**	**3,29**	**0,85**	**6,42**
Bocageopsis canescens	42	0,60	0,44	0,18	1,22
Duguetia cf. *argentea*	2	0,03	0,22	0,01	0,26
Duguetia sp.	1	0,01	0,22	0,00	0,24
Duguetia uniflora	15	0,21	1,10	0,08	1,39
Guatteria meliodora	1	0,01	0,22	0,02	0,25
Guatteria schomburgkiana	7	0,10	0,44	0,07	0,61
Pseudoxandra cf. *Xilopioides*	1	0,01	0,22	0,01	0,25
Pseudoxandra lucida	41	0,59	0,44	0,16	1,18
Xylopia aromatica	24	0,34	0,44	0,21	0,99
Xylopia parviflora	24	0,34	0,22	0,10	0,67
Xylopia sp.	2	0,03	0,22	0,01	0,26
Apocynaceae	**59**	**0,84**	**2,82**	**0,53**	**4,19**
Aspidosperma nitidum	5	0,07	0,44	0,03	0,54
Aspidosperma spruceanum	22	0,31	0,44	0,13	0,89
Couma utilis	26	0,37	0,44	0,34	1,15
Himatanthus attenuathus	4	0,06	0,66	0,01	0,73
Malouetia cf. *tamaquarina*	2	0,03	0,44	0,01	0,48
Arecaceae	**10**	**0,14**	**1**	**0,11**	**1**
Astrocaryum jauari	8	0,11	0,44	0,06	0,61
Attalea maripa	2	0,03	0,22	0,05	0,30
Bignoniaceae	**23**	**0,33**	**2,35**	**0,40**	**3,08**
Handroanthus barbatus	23	0,33	1,10	0,40	1,83
Burseraceae	**61**	**0,87**	**1,41**	**0,27**	**2,55**
Protium sp.	60	0,86	0,66	0,26	1,78
Trattinnickia cf. *rhoifolia*	1	0,01	0,22	0,01	0,24
Calophyllaceae	**94**	**1,34**	**3,29**	**1,39**	**6,02**
Calophyllum brasiliense	36	0,51	1,10	0,85	2,47
Caraipa grandifolia	15	0,21	0,66	0,05	0,92
Caraipa odorata	5	0,07	0,22	0,03	0,32
Marila laxiflora	38	0,54	0,22	0,46	1,23
Caryocaraceae	**4**	**0,06**	**0,94**	**0,33**	**1,33**
Caryocar glabrum	4	0,06	0,44	0,33	0,83
Chrysobalanaceae	**244**	**3,49**	**4,23**	**1,95**	**9,66**
Couepia cf. *guianensis*	2	0,03	0,22	0,01	0,26
Couepia paraensis	20	0,29	0,88	0,10	1,27
Hirtella bicornis	49	0,70	0,22	0,45	1,37
Licania apetala	31	0,44	1,32	0,27	2,03
Licania cf. *adolphoduckei*	16	0,23	0,22	0,09	0,54
Licania heteromorpha	57	0,82	0,88	0,37	2,07
Licania hypoleuca	15	0,21	0,22	0,12	0,55
Licania micrantha	6	0,09	0,22	0,08	0,39
Licania octandra	6	0,09	0,44	0,04	0,57
Licania sp. 1	36	0,51	0,44	0,31	1,26
Licania sp. 2	6	0,09	0,22	0,10	0,41

Appendix B (cont'd) - List of families and species recorded and their phytosociological

parameter values.

Family/Species	Abund.	Relative Density Relative	Relative Frequency Relative	Dom. Relative	IVI

Clusiaceae	**11**	**0,16**	**1,88**	**0,07**	**2,11**
Garcinia madruno	2	0,03	0,22	0,01	0,26
Tovomita spruceana	9	0,13	0,66	0,06	0,85
Combretaceae	**124**	**1,77**	**3,76**	**2,21**	**7,74**
Buchenavia ochroprumna	117	1,67	0,88	2,08	4,63
Buchenavia sericocarpa	3	0,04	0,22	0,05	0,31
Buchenavia sp.	1	0,01	0,22	0,00	0,24
Terminalia sp.	3	0,04	0,44	0,08	0,57
Ebenaceae	**68**	**0,97**	**2,35**	**0,46**	**3,78**
Diospyros cf. *guianensis*	2	0,03	0,22	0,03	0,27
Diospyros cf. *vestita*	65	0,93	0,66	0,43	2,02
Diospyros sp.	1	0,01	0,22	0,00	0,24
Elaeocarpaceae	**9**	**0,13**	**0,94**	**0,07**	**1,14**
Sloanea cf..*floribunda*	9	0,13	0,44	0,07	0,64
Erythroxylaceae	**166**	**2,37**	**2,82**	**0,95**	**6,14**
Erythroxylum spruceanum	166	2,37	1,32	0,95	4,64
Euphorbiaceae	**386**	**5,52**	**4,23**	**4,98**	**14,7**
Alchornea discolor	1	0,01	0,22	0,00	0,24
Alchorneopsis floribunda	20	0,29	0,22	0,18	0,69
Hevea brasiliensis	5	0,07	0,22	0,12	0,41
Hevea cf. *benthamiana*	3	0,04	0,22	0,02	0,28
Hevea cf. *microphylla*	2	0,03	0,22	0,01	0,25
Hevea spruceana	179	2,56	0,66	3,28	6,50
Mabea nitida	123	1,76	1,32	1,07	4,15
Maprounea amazonica	53	0,76	1,76	0,30	2,82
Fabaceae	**1956**	**27,97**	**4,69**	**47,24**	**79,9**
Acosmium nitens	54	0,77	1,32	0,71	2,80
Aldina latifolia	144	2,06	1,32	6,14	9,52
Andira cf. *unifoliolata*	3	0,04	0,44	0,03	0,51
Andira inermis	3	0,04	0,22	0,43	0,69
Andira micrantha	25	0,36	0,66	0,80	1,82
Campsiandra cf. *angustifolia*	16	0,23	0,44	0,33	1,00
Clathrotropis nitida	23	0,33	0,88	0,31	1,51
Crudia amazonica	25	0,36	0,44	0,16	0,95
Cynometra sp.	12	0,17	0,88	0,17	1,22
Dialium guianense	2	0,03	0,22	0,06	0,31
Dicorynia cf. *paraensis*	1	0,01	0,22	0,08	0,32
Diplotropis cf. *martiusii*	5	0,07	0,22	0,13	0,42
Diplotropis purpurea	2	0,03	0,22	0,04	0,29
Elizabetha speciosa	5	0,07	0,22	0,02	0,31
Heterostemon mimosoides	6	0,09	0,66	0,03	0,77
Hydrochorea marginata	98	1,40	1,32	2,58	5,30
Inga ulei	3	0,04	0,22	0,01	0,27
Limadendron amazonica	11	0,16	1,10	0,04	1,30
Macrolobium acaciifolium	306	4,38	1,76	8,68	14,81

Appendix B (cont'd) - List of families and species recorded and their phytosociological parameter values.

Family/Species	**Abund.**	**Relative Density Relative**	**Relative Frequency Relative**	**Dom. Relative**	**IVI**
Fabaceae	**1956**	**27,97**	**4,69**	**47,24**	**79,9**
Macrolobium multijugum	10	0,14	0,22	0,26	0,63
Ormosia excelsa	159	2,27	0,88	3,71	6,86
Ormosia sp.	1	0,01	0,22	0,02	0,25
Parkia cf. *panurensis*	2	0,03	0,22	0,04	0,28
Parkia discolor	29	0,41	0,88	0,39	1,69
Parkia nitida	1	0,01	0,22	0,04	0,27
Parkia pendula	1	0,01	0,22	0,02	0,26

Peltogyne cf. *catingae*	27	0,39	0,44	0,37	1,20
Stryphnodendron sp.	2	0,03	0,22	0,04	0,28
Swartzia cf. *pendula*	5	0,07	0,44	0,03	0,54
Swartzia cuspidata	10	0,14	0,44	0,08	0,67
Swartzia ingifolia	13	0,19	0,44	0,17	0,80
Swartzia laevicarpa	135	1,93	0,66	1,01	3,60
Swartzia polyphylla	239	3,42	1,54	7,93	12,89
Tachigali myrmecophila	1	0,01	0,22	0,00	0,24
Tachigali sp.	532	7,61	1,76	11,54	20,91
Taralea oppositifolia	1	0,01	0,22	0,03	0,26
Vatairea guianensis	19	0,27	0,88	0,58	1,73
Zygia cf. *longifolia*	24	0,34	0,22	0,24	0,80
Zygia sp.	1	0,01	0,22	0,00	0,24
Goupiaceae	**20**	**0,29**	**1,41**	**0,20**	**1,9**
Goupia glabra	20	0,29	0,66	0,20	1,15
Humiriaceae	**70**	**1,00**	**2,82**	**1,18**	**5**
Humiriastrum cuspidatum	2	0,03	0,22	0,06	0,31
Sacoglottis ceratocarpa	10	0,14	0,22	0,12	0,48
Sacoglottis guianensis	33	0,47	1,32	0,79	2,58
Vantanea guianensis	25	0,36	0,22	0,22	0,79
Hypericaceae	**2**	**0,03**	**0,47**	**0,03**	**0,53**
Vismia cayennensis	2	0,03	0,22	0,03	0,28
Icacinaceae	**25**	**0,36**	**0,94**	**0,51**	**1,81**
Emmotum cf. *acuminatum*	25	0,36	0,44	0,51	1,31
Lauraceae	**156**	**2,23**	**3,76**	**1,77**	**7,75**
Aniba affinis	33	0,47	1,54	0,64	2,65
Aniba parviflora	2	0,03	0,44	0,02	0,49
Endlicheria anomala	6	0,09	0,22	0,03	0,34
Endlicheria levelii	64	0,92	0,88	0,41	2,20
Endlicheria ruforamula	3	0,04	0,22	0,01	0,28
Nectandra amazonum	45	0,64	1,10	0,64	2,38
Ocotea sp.	3	0,04	0,22	0,01	0,28
Lecythidaceae	**399**	**5,71**	**4,69**	**10,60**	**21**
Bertholletia excelsa	1	0,01	0,22	0,04	0,27
Couratari tenuicarpa	14	0,20	0,44	0,24	0,88
Eschweilera cf. *albiflora*	3	0,04	0,22	0,01	0,27
Eschweilera cf. *ovalifolia*	9	0,13	0,22	0,05	0,40

Appendix B (cont'd) - List of families and species recorded and their phytosociological parameter values.

Family/Species	Abund.	Relative Density Relative	Relative Frequency Relative	Dom. Relative	IVI
Lecythidaceae	**399**	**5,71**	**4,69**	**10,60**	**21**
Eschweilera cf. *parvifolia*	18	0,26	0,88	0,21	1,35
Eschweilera sp.	13	0,19	0,44	0,09	0,72
Eschweilera sp. 1	16	0,23	0,22	0,13	0,58
Eschweilera sp. 2	3	0,04	0,22	0,04	0,30
Eschweilera sp. 3	22	0,31	0,22	0,13	0,66
Eschweilera tenuifolia	257	3,68	1,32	9,09	14,09
Lecythis chartacea	43	0,61	1,10	0,57	2,29
Malpighiaceae	**354**	**5,06**	**4,69**	**2,33**	**12,1**
Acmanthera latifolia	2	0,03	0,44	0,01	0,47
Burdachia sp.	269	3,85	1,54	1,76	7,14
Byrsonima cf. *punctulata*	54	0,77	0,88	0,30	1,95
Byrsonima cf. *schomburgkiana*	8	0,11	0,66	0,04	0,81
Byrsonima incarnata	19	0,27	0,22	0,21	0,70
Byrsonima sp.	1	0,01	0,22	0,01	0,25
Glandonia cf. *williamsii*	1	0,01	0,22	0,01	0,24

Malvaceae	**31**	**0,44**	**2,35**	**0,31**	**3,1**
Lueheopsis cf. *althaeiflora*	2	0,03	0,22	0,01	0,26
Mollia cf. *lepidota*	12	0,17	0,44	0,17	0,78
Mollia cf. *speciosa*	17	0,24	0,44	0,13	0,81
Melastomataceae	**75**	**1,07**	**1,88**	**0,61**	**3,56**
Bellucia spruceana	2	0,03	0,22	0,01	0,26
Miconia cf. *pubipetala*	5	0,07	0,22	0,02	0,31
Miconia poeppigii	18	0,26	0,22	0,22	0,70
Mouriri cf. *apiranga*	1	0,01	0,22	0,01	0,24
Mouriri cf. *grandiflora*	1	0,01	0,22	0,00	0,24
Mouriri cf. *myrtifolia*	9	0,13	0,22	0,13	0,48
Mouriri sp.	39	0,56	0,44	0,22	1,22
Meliaceae	**2**	**0,03**	**0,94**	**0,01**	**0,98**
Trichilia rubra	2	0,03	0,44	0,01	0,48
Moraceae	**7**	**0,10**	**0,94**	**0,04**	**1,08**
Brosimum guianense	6	0,09	0,22	0,03	0,34
Ficus maxima	1	0,01	0,22	0,01	0,24
Myristicaceae	**10**	**0,14**	**1,88**	**0,03**	**2,05**
Virola cf. *sebifera*	1	0,01	0,22	0,00	0,24
Virola elongata	9	0,13	0,66	0,03	0,82
Myrtaceae	**194**	**2,77**	**3,76**	**1,01**	**7,54**
Calyptranthes cf. *nigrescens*	77	1,10	1,32	0,32	2,74
Calyptranthes cf. *spruceana*	51	0,73	0,66	0,27	1,66
Calyptranthes sp.	16	0,23	0,66	0,11	1,00
Eugenia cf. *lambertiana*	1	0,01	0,22	0,00	0,24
Eugenia gomesiana	8	0,11	0,22	0,04	0,37
Eugenia sp. 1	1	0,01	0,22	0,01	0,24
Eugenia sp. 2	1	0,01	0,22	0,01	0,24
Marlierea cf. *bipennis*	4	0,06	0,22	0,02	0,29

Appendix B (cont'd) - List of families and species recorded and their phytosociological parameter values.

Family/Species	Abund.	Relative Density Relative	Relative Frequency Relative	Dom. Relative	IVI
Myrtaceae	**194**	**2,77**	**3,76**	**1,01**	**7,54**
Marlierea cf. *umbraticola*	3	0,04	0,22	0,03	0,30
Marlierea ensiformis	3	0,04	0,22	0,02	0,28
Myrcia cf. *gigas*	29	0,41	1,10	0,18	1,70
Ochnaceae	**320**	**4,58**	**2,35**	**1,73**	**8,65**
Blastemanthus cf. *gemmiflorus*	6	0,09	0,22	0,02	0,33
Elvasia quinqueloba	304	4,35	1,10	1,67	7,12
Ouratea cf. *chrysopetala*	1	0,01	0,22	0,00	0,24
Ouratea cf. *pisiformis*	9	0,13	0,22	0,04	0,38
Olacaceae	**10**	**0,14**	**1,41**	**0,10**	**1,65**
Chaunochiton kappleri	4	0,06	0,22	0,02	0,30
Chaunochiton loranthoides	6	0,09	0,44	0,08	0,60
Pentaphylacaceae	**128**	**1,83**	**4,23**	**1,17**	**7,23**
Ternstroemia candolleana	128	1,83	1,98	1,17	4,98
Peraceae	**8**	**0,11**	**0,94**	**0,08**	**1,13**
Pera distichophylla	8	0,11	0,44	0,08	0,63
Phyllanthaceae	**518**	**7,41**	**3,29**	**5,15**	**15,9**
Amanoa oblongifolia	497	7,11	1,32	5,03	13,46
Discocarpus essequeboensis	13	0,19	0,66	0,06	0,91
Richeria grandis	8	0,11	0,22	0,06	0,40
Polygonaceae	**1**	**0,01**	**0,47**	**0,01**	**0,49**
Ruprechtia tangarana	1	0,01	0,22	0,01	0,24
Primulaceae	**7**	**0,10**	**1,88**	**0,04**	**2,02**

Cybianthus cf. *spicatus*	7	0,10	0,88	0,04	1,02
Proteaceae	**1**	**0,01**	**0,47**	**0,01**	**0,49**
Panopsis cf. *sessilifolia*	1	0,01	0,22	0,01	0,24
Rubiaceae	**264**	**3,78**	**3,29**	**1,94**	**9**
Duroia cf. *velutina*	254	3,63	1,32	1,90	6,85
Duroia sp.	10	0,14	0,22	0,04	0,41
Salicaceae	**119**	**1,70**	**2,35**	**0,91**	**4,96**
Homalium racemosum	78	1,12	0,88	0,72	2,71
Laetia suaveolens	41	0,59	0,66	0,20	1,44
Sapotaceae	**821**	**11,74**	**4,23**	**7,40**	**23,4**
Ecclinusa lancifolia	16	0,23	0,44	0,57	1,24
Ecclinusa sp.	1	0,01	0,22	0,00	0,24
Elaeoluma glabrescens	5	0,07	0,22	0,03	0,33
Elaeoluma nuda	7	0,10	0,44	0,15	0,69
Elaeoluma schomburgkiana	13	0,19	0,22	0,08	0,49
Manilkara inundata	25	0,36	0,44	0,20	1,00
Micropholis cf. *guyanensis*	3	0,04	0,22	0,02	0,29
Micropholis cf. *venulosa*	2	0,03	0,22	0,03	0,27
Micropholis guyanensis	61	0,87	0,88	0,39	2,14
Micropholis humboldtiana	154	2,20	1,10	2,13	5,43
Pouteria campanulata	5	0,07	0,44	0,03	0,54
Pouteria cf. *gardneri*	1	0,01	0,22	0,01	0,24

Appendix B (cont'd) - List of families and species recorded and their phytosociological

parameter values.

Family/Species	Abund.	Relative Density Relative	Relative Frequency Relative	Dom. Relative	IVI
Sapotaceae	**821**	**11,74**	**4,23**	**7,40**	**23,4**
Pouteria elegans	499	7,14	1,76	3,59	12,49
Pouteria pachyphylla	9	0,13	0,22	0,05	0,39
Pradosia aff. *Schomburgkiana*	20	0,29	0,22	0,12	0,63
Simaroubaceae	**40**	**0,57**	**3,29**	**0,45**	**4,3**
Simaba cedron	1	0,01	0,22	0,00	0,24
Simaba cf. *obovata*	34	0,49	0,88	0,34	1,71
Simaba orinocensis	5	0,07	0,44	0,10	0,61
Vochysiaceae	**35**	**0,50**	**1,41**	**0,57**	**2,48**
Erisma calcaratum	1	0,01	0,22	0,00	0,24
Qualea sp.	14	0,20	0,44	0,17	0,81
Ruizterania retusa	11	0,16	0,44	0,34	0,94
Vochysia sp.	9	0,13	0,44	0,06	0,63

Appendix C - Classification and occurrence of species according to flood tolerance.

Flood status	Low	Intermediate	High	Very high	Species tolerance
Species/Flooding interval (m)	**0-1 1-2 2-3**	**3-4 4-55-6**	**6-7 7-8 8-9**	**10-11 >11**	**the flood**
Alchomeopsis floribunda	1082				Low
Andira cf. *unifoliolata*	21				Low
Andira inermis	21				Low
Andira micrantha	9123	1			Low
Aniba parviflora	11				Low
Aspidosperma spruceanum	4108				Low
Attalea maripa	11				Low
Blastemanthus cf. gemmiflorus	123				Low
Bocageopsis canescens	4326				Low
Brosimum guianense	411				Low
Buchenavia sericocarpa	3				Low
Byrsonima cf. schomburgkiana	24	2			Low
Caryocar glabrum	22				Low

Chaunochiton kappleri	22				Low
Clathrotropis nitida	379	22			Low
Couma utilis	521				Low
Couratari tenuicarpa	115	7			Low
Dialium guianense	2				Low
Diplotropis cf. *martiusii*	32				Low
Diplotropis purpurea	2				Low
Elaeoluma schomburgkiana	58				Low
Emmotum cf. acuminatum	619				Low
Eschweilera cf. *albiflora*	3				Low
Eschweilera cf. *ovalifolia*	45				Low
Eschweilera sp.	211				Low
Eschweilera sp. 1	574				Low
Eschweilera sp. 2	12				Low
Eschweilera sp. 3	2109	1			Low
Garcinia madruno	11				Low
Goupia glabra	6122				Low
Guatteria schomburgkiana	25				Low
Hevea cf. *microphylla*	11				Low
Hirtella bicornis	1039				Low
Inga ulei	3				Low
Licania cf. *adolphoduckei*	3111	1			Low
Licania hypoleuca	681				Low
Licania micrantha	113	1			Low
Licania octandra	51				Low
Licania sp. 2	232				Low
Lueheopsis cf. *althaeiflora*	2				Low
Manilkara inundata	5182				Low
Marila laxiflora	533				Low
Marlierea cf. *umbraticola*	11	1			Low
Miconia cf. *pubipetala*	41				Low

Appendix C (cont'd) - Classification and occurrence of species according to flood tolerance.

Flood status	Low	Intermediate	High	Very high	Species tolerance
Species/Flooding interval (m)	**0-1 1-2 2-3**	**3-4 4-55-6**	**6-7 7-8 8-9**	**10-11 >11**	**the flood**
Miconia poeppigii	6111				Low
Micropholis cf. *guyanensis*	2	1			Low
Mollia cf. *lepidota*	56	1			Low
Mouriri cf. *myrtifolia*	342				Low
Mouriri sp.	13176	3			Low
Ocotea sp.	3				Low
Ouratea cf. *pisiformis*	54				Low
Parkia cf. *panurensis*	2				Low
Pera distichophylla	17				Low
Pouteria pachyphylla	9				Low
Pradosia aff. *schomburgkiana*	3105	2			Low
Protium sp.	1842				Low
Qualea sp.	212				Low
Richeria grandis	44				Low
Ruizterania retusa	191				Low
Sloanea cf. *floribunda*	45				Low
Swartzia cf. *pendula*	22	1			Low
Swartzia cuspidata	631				Low
Swartzia ingifolia	310				Low
Vantanea guianensis	817				Low
Vismia cayennensis	11				Low
Vochysia sp.	63				Low
Xylopia aromatica	2211				Low
Xylopia parviflora	1113				Low
Xylopia sp.	11				Low
Zygia cf. *longifolia*	517	2			Low
Acmanthera latifolia		11			Intermediate
Aldina latifolia	103324	191840			Intermediate
Aniba affinis	21	565	131		Intermediate
Aspidosperma nitidum	12	2			Intermediate

Astrocaryum jauari		42	2		Intermediate
Bellucia spruceana		2			Intermediate
Byrsonima cf. *punctulata*	518	1912			Intermediate
Byrsonima incarnata		575	2		Intermediate
Calyptranthes cf. *nigrescens*	341	12930	216		Intermediate
Calyptranthes cf. *spruceana*	2	3541			Intermediate
Calyptranthes sp.	1	2211			Intermediate
Campsiandra cf. *angustifolia*	3	751			Intermediate
Caraipa grandifolia		1112	1		Intermediate
Caraipa odorata	3	2			Intermediate
Chaunochiton loranthoides		22	2		Intermediate
Couepia cf. *guianensis*			2		Intermediate
Crudia amazonica		595	24		Intermediate
Cybianthus cf. *spicatus*		32	11		Intermediate

Appendix C (cont'd) - Classification of species according to flood tolerance.

Flood status	Low	Intermediate	High	Very high	Species tolerance
Species/Flooding interval (m)	**0-1 1-2 2-3**	**3-4 4-55-6**	**6-7 7-8 8-9**	**10-11 >11**	**the flood**
Cynometra sp.	2	27	1		Intermediate
Diospyros cf. *guianensis*		11			Intermediate
Diospyros cf. *vestita*		1855	1		Intermediate
Discocarpus essequeboensis	1	111			Intermediate
Duguetia cf. *argentea*			2		Intermediate
Duguetia uniflora	341	115			Intermediate
Duroia cf. *velutina*	2712	571547	9519		Intermediate
Duroia sp.		73			Intermediate
Ecclinusa lancifolia		115			Intermediate
Elaeoluma nuda	1	3	3		Intermediate
Elizabetha speciosa	1	4			Intermediate
Elvasia quinqueloba	117450	117186	1		Intermediate
Endlicheria anomala		123			Intermediate
Endlicheria levelii		51239	8		Intermediate
Endlicheria ruforamula		21			Intermediate
Erythroxylum spruceanum	33	727120	7		Intermediate
Eschweilera cf. *parvifolia*	4	113			Intermediate
Eugenia gomesiana		8			Intermediate
Handroanthus barbatus	11	5610			Intermediate
Heterostemon mimosoides	11	3	1		Intermediate
Hevea brasiliensis	11	21			Intermediate
Hevea cf. *benthamiana*		3			Intermediate
Hevea spruceana	2	384139	1940		Intermediate
Himatanthus attenuathus	1	11	1		Intermediate
Homalium racemosum		26327	373		Intermediate
Humiriastrum cuspidatum		1	1		Intermediate
Hydrochorea marginata	6	431810	5142		Intermediate
Laetia suaveolens	7	293	2		Intermediate
Lecythis chartacea	7185	274			Intermediate
Licania heteromorpha	2138	331			Intermediate
Licania sp. 1		729			Intermediate
Limadendron amazonica	14	13	2		Intermediate
Mabea nitida	12	57612	1719		Intermediate
Malouetia cf. *tamaquarina*		1	1		Intermediate
Maprounea amazonica	8	18511	821		Intermediate
Marlierea cf. *bipennis*		31			Intermediate
Marlierea ensiformis		12			Intermediate
Micropholis cf. *venulosa*	1	1			Intermediate
Micropholis guyanensis	11	2755	832		Intermediate
Micropholis humboldtiana	233	9750	746		Intermediate
Mollia cf. *speciosa*	1	412			Intermediate
Myrcia cf. *gigas*		1219	52		Intermediate
Nectandra amazonum	1	7414	181		Intermediate
Parkia discolor	63	110	9		Intermediate

Appendix C (cont'd) - Classification of species according to flood tolerance.

Flood status	Low	Intermediate	High	Very high	Species tolerance
Species/Flooding interval (m)	**0-1 1-2 2-3**	**3-4 4-5 5-6**	**6-7 7-8 8-9**	**10-11 >11**	**the flood**

Peltogyne cf. *catingae*	6192				Intermediate
Pouteria campanulata		12	2		Intermediate
Pouteria elegans	104016	20136 144	83482		Intermediate
Pseudoxandra lucida		2619	212		Intermediate
Sacoglottis ceratocarpa		10			Intermediate
Sacoglottis guianensis	954	924			Intermediate
Simaba cf. *obovata*	124	1017			Intermediate
Stryphnodendron sp.		11			Intermediate
Swartzia laevicarpa	13	135663			Intermediate
Swartzia polyphylla	379	2361135	1		Intermediate
Terminalia sp.		12			Intermediate
Ternstroemia candolleana	389	2165	139		Intermediate
Tovomita spruceana		135			Intermediate
Trichilia rubra		2			Intermediate
Vatairea guianensis	14	734			Intermediate
Virola elongata		117			Intermediate
Acosmium nitens	4	1133	9177		High
Buchenavia ochroprumna		634	98510		High
Couepia paraensis	1	11	584		High
Elaeoluma glabrescens			5		High
Licania apetala	123	312	163		High
Macrolobium acaciifolium	3	414852	2011129	2	High
Macrolobium multijugum			91		High
Ormosia excelsa	8	32137	146718		High
Simaba orinocensis		1	31		High
Eschweilera tenuifolia	11	662	6496	7 1631	Very high
Amanoa oblongifolia	33	1713619	3617131		Irregular
Burdachia sp.	22	1031829	34556	2	Irregular
Calophyllum brasiliense	1072	1114	1		Irregular
Tachigali sp.	238578	7169135	5616		Irregular

Appendix D - Relative abundance per species along the flood gradient for the 60 most abundant species.

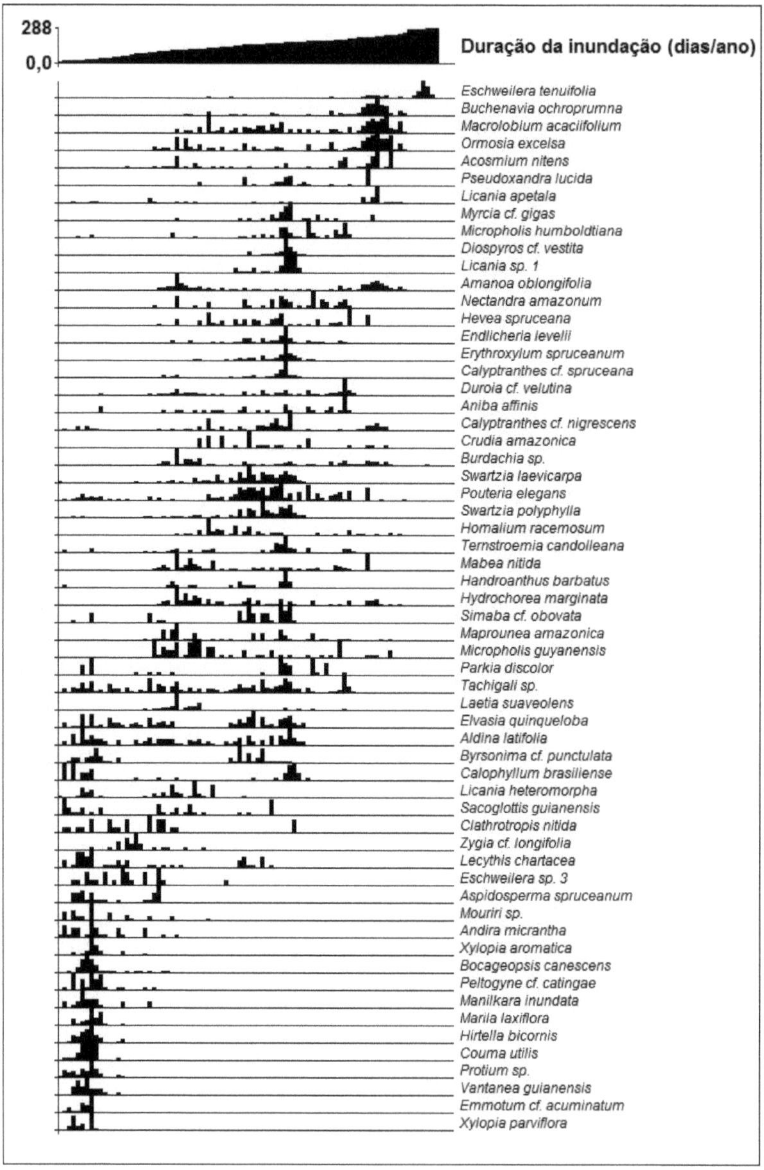

9. ANNEXES

Annex A - Location map of the study plots.

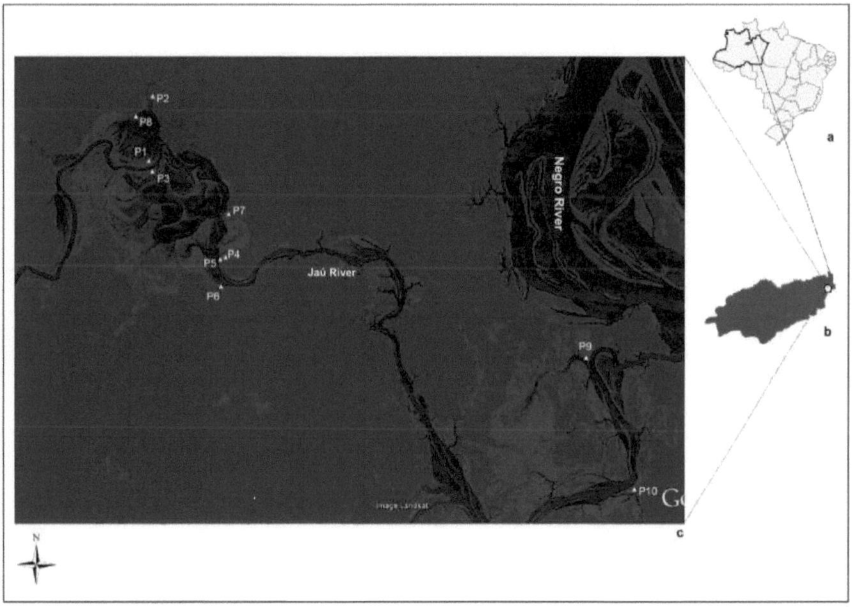

Annex B - Values of graphic variables by quadrant (Source: Embrapa Amazônia West).

Quadrant	pH (H_2O)	C	M.O.	N	P	K	Na	Ca	Mg	Fe	Zn	Mn	Cu	Sand	Silt	Clay
			(g/kg)		(mg/dm³)			(cmolc/dm³)			(mg/dm³)				(g/kg)	
1ABCD	4,28	59,5	102,4	2,9	10,7	60	10	0,08	0,17	45	1,16	0,94	0,25	18,4	554,6	427
1EFGH	4,44	29	49,8	2,4	10,1	39	27	0,06	0,13	53	0,64	0,63	0,28	26,7	514,3	459
1IJKL	4,4	70,3	120,9	5,6	22,4	86	34	0,12	0,28	21	1,26	1,62	0,44	143,5	606,5	250
1MNOP	4,3	38,3	65,9	2,9	8,8	50	30	0,06	0,19	57	0,98	2,37	0,28	18,8	611,1	370
2ABCD	4,39	31,6	54,3	2,2	8,3	26	6	0,07	0,1	25	0,66	0,52	0,26	63,7	406,8	529,5
2EFGH	4,45	40,7	70	3,2	7,7	30	5	0,05	0,09	14	0,58	0,61	0,25	77,8	442,7	479,5
2IJKL	4	14,3	24,6	1,3	10,2	13	3	0,04	0,07	15	0,39	0,61	0,11	365,7	352,3	282
2MNOP	4,33	18,3	31,5	1,6	9,8	21	4	0,05	0,08	39	0,45	0,39	0,2	397,7	420,8	181,5
3ABCD	4,1	18,7	32,1	1,7	11,4	39	5	0,06	0,11	83	0,62	1,05	0,19	106,3	529,2	364,5
3EFGH	4,44	33	56,8	4	19,5	52	9	0,06	0,14	76	0,97	1,11	0,3	52	553,5	394,5
3IJKL	4,35	54,6	94	4,4	18,4	67	7	0,07	0,16	47	1,01	1,44	0,31	44,3	613,7	342
3MNOP	4,03	38,6	66,4	2,7	17,8	95	13	0,11	0,19	39	0,76	1,14	0,16	82,5	620,5	297
4ABCD	4,42	24	41,3	2,2	8,2	40	5	0,08	0,14	61	0,75	1,47	0,14	140,4	493,1	366,5
4EFGH	4,15	31	53,3	2,2	17,9	60	4	0,09	0,18	28	0,91	1,47	0,16	131,8	558,2	310
4IJKL	4,56	22,6	38,9	1,9	10,1	61	5	0,08	0,13	117	0,79	1,16	0,26	225	533,5	241,5
4MNOP	4,41	30,7	52,8	2,5	15,8	77	4	0,06	0,23	105	0,88	1,35	0,28	144,9	606,6	248,5
5ABCD	4,17	14,2	24,4	1,3	16,5	48	4	0,08	0,17	47	0,61	1,06	0,24	214,8	562,7	222,5
5EFGH	4,28	26,2	45,1	2	8,1	58	7	0,08	0,17	138	0,93	1,65	0,25	154,2	628,8	217
5IJKL	4,26	29,5	50,8	2,5	13,8	77	6	0,1	0,22	104	0,83	1,83	0,31	123,2	610,8	266
5MNOP	4,43	21,6	37,2	1,7	8,4	76	5	0,08	0,17	33	0,64	0,88	0,23	132,6	699,4	168
6ABCD	4,61	16,5	28,4	1,4	11,5	35	7	0,07	0,09	135	0,41	0,5	0,26	592	280	128
6EFGH	4,72	31,1	53,6	2,4	14,3	36	6	0,07	0,11	51	0,75	0,52	0,3	210,1	515,9	274
6IJKL	4,64	32,2	55,5	2,2	12,3	39	6	0,06	0,11	47	1,51	0,65	0,29	487,2	302,8	210
6MNOP	4,68	29,2	50,2	1,9	19,2	52	9	0,08	0,13	69	0,76	0,48	0,26	536,5	323,5	140
7ABCD	3,87	4,2	7,2	0,5	7,4	14	3	0,04	0,06	289	0,25	0,26	0,22	467,5	428,5	104
7EFGH	4	6,6	11,3	0,7	7,9	14	3	0,04	0,06	250	0,29	0,29	0,31	414,1	498,4	87,5
7IJKL	3,95	8,7	14,9	0,9	13,3	23	4	0,05	0,08	170	0,56	0,66	0,21	295,2	609,8	95
7MNOP	3,66	7,5	12,9	0,7	9,8	12	2	0,05	0,06	282	0,48	0,28	0,22	338,3	567,7	94
9ABCD	4,21	22,6	38,9	1,7	7,2	27	6	0,04	0,07	81	0,6	0,53	0,24	62,6	672,4	265
9EFGH	4,11	17,3	29,8	1,6	8,6	26	6	0,05	0,08	49	0,73	0,57	0,24	49,9	654,6	295,5
9IJKL	4,17	30,4	52,3	2,8	8,6	39	6	0,06	0,1	57	0,61	0,62	0,28	58,5	645	296,5
9MNOP	4,21	16,5	28,3	1,8	5,5	18	4	0,04	0,06	103	0,47	0,31	0,19	50,5	568	381,5
10ABCD	4,18	20,2	34,7	1,8	7,5	30	5	0,05	0,08	147	0,54	0,52	0,24	239,5	524	236,5
10EFGH	4,18	22,5	38,8	2	9,5	36	4	0,06	0,11	134	0,92	0,61	0,29	136,1	630,4	233,5
10IJKL	4,59	17,2	29,6	1,2	5,4	22	5	0,04	0,07	236	0,48	0,47	0,24	128,9	651,1	220
10MNOP	4,61	18,5	31,7	1,6	9,3	22	3	0,06	0,08	115	0,66	0,35	0,26	134,5	644,6	221

MIX
Papier aus verantwortungsvollen Quellen
Paper from responsible sources
FSC® C105338

Printed by Books on Demand GmbH, Norderstedt / Germany